Research Advances in Network Technologies

In the current digital era, information is an important asset for our daily lives as well as for small- and large-scale businesses. The network technologies are the main enablers that connect the computing devices and resources together to collect, process, and share vital information locally as well as globally. The network technologies provide efficient, flexible, and seamless communication while maximizing productivity and resources for our day-to-day lives and business operations. For all its importance, this domain has evolved considerably in the last few decades, from the traditional wired networks to Bluetooth, infrared waves, micro-waves, radio-waves, and satellite networks. Nowadays, network technologies are not only restricted to computer laboratories, offices, or homes; many other diverse areas have been witnessed where network technologies are being used based on the applications and needs, such as vehicular ad hoc networks, underwater networks, and the Internet of Things.

Along with the hardware-based and physical network technologies, a lot of research has been carried out by researchers from academia and industry to develop emerging software-based network technologies, such as network software architectures, middleware, and protocol stacks. The software-based network technologies become the main driving force behind the paradigm shift in this domain and have invented many new network technologies such as grid computing, cloud computing, fog computing, edge computing, software-defined networks, and content-centric networks. On the other hand, a lot of efforts have been made in cellular network technologies to improve the user experience, and as a consequence, emerging cellular network technologies like long-term evolution (LTE), Voice over Long-Term Evolution (VoLTE), and 5G have been invented. Due to its demand and importance in present and future scenarios, numerous efforts have been made in the networking domain by the researchers, a lot of work is still ongoing, and many more possibilities have yet to be explored. Therefore, there is a need to keep track of advancements related to network technologies and further investigate several ongoing research challenges for the ease of users. With this goal in mind, **Research Advances in Network Technologies** presents the most recent and notable research on network technologies.

Research Advances in Network Technologies

Edited by
Anshul Verma, Pradeepika Verma,
Kiran Kumar Pattanaik, and Lalit Garg

CRC Press
Taylor & Francis Group
Boca Raton London New York

CRC Press is an imprint of the
Taylor & Francis Group, an **informa** business

MATLAB® is a trademark of The MathWorks, Inc. and is used with permission. The MathWorks does not warrant the accuracy of the text or exercises in this book. This book's use or discussion of MATLAB® software or related products does not constitute endorsement or sponsorship by The MathWorks of a particular pedagogical approach or particular use of the MATLAB® software

First edition published 2023
by CRC Press
6000 Broken Sound Parkway NW, Suite 300, Boca Raton, FL 33487-2742

and by CRC Press
4 Park Square, Milton Park, Abingdon, Oxon, OX14 4RN

CRC Press is an imprint of Taylor & Francis Group, LLC

© 2023 selection and editorial matter, Anshul Verma, Pradeepika Verma, Kiran Kumar Pattanaik and Lalit Garg; individual chapters, the contributors

Library of Congress Cataloging-in-Publication Data
Names: Verma, Anshul, editor. | Verma, Pradeepika, editor. | Pattanaik, Kiran Kumar, editor. | Garg, Lalit, 1977- editor.
Title: Research advances in network technologies / edited by Anshul Verma, Pradeepika Verma, Kiran Kumar Pattanaik, and Lalit Garg.
Description: First edition. |
Boca Raton : CRC Press, 2023. |
Includes bibliographical references and index.
Identifiers: LCCN 2022040483 (print) | LCCN 2022040484 (ebook) | ISBN 9781032340487 (hbk) | ISBN 9781032340494 (pbk) | ISBN 9781003320333 (ebk)
Subjects: LCSH: Computer networks—Technological innovations. | Computer networks—Research.
Classification: LCC TK5105.5.R4484 2023 (print) | LCC TK5105.5 (ebook) | DDC 004.6—dc23/
eng/20221109
LC record available at https://lccn.loc.gov/2022040483
LC ebook record available at https://lccn.loc.gov/2022040484

ISBN: 9781032340487 (hbk)
ISBN: 9781032340494 (pbk)
ISBN: 9781003320333 (ebk)

DOI: 10.1201/9781003320333

Typeset in Times
by codeMantra

Contents

Preface

OVERVIEW AND GOALS

In the current digital era, information is an important asset for our daily life as well as for small- and large-scale businesses. The network technologies are the main enablers that connect the computing devices and resources together to collect, process, and share the vital information locally as well as globally. The network technologies provide efficient, flexible, and seamless communication while maximizing productivity and resources for our day-to-day life and business operations. For all these importance, this domain has been evolved drastically in the last few decades from the traditional wired networks to the Bluetooth, infrared, micro-waves, radio-waves, and satellite networks. Nowadays, network technologies are not only restricted to computer laboratories, offices, or homes, but many other diverse areas have been witnessed where network technologies are being used based on the applications and needs such as vehicular ad hoc networks, underwater networks, and Internet of things.

Along with the hardware-based and physical network technologies, a lot of research works have been carried out by the researchers from academia and industry to develop the emerging software-based network technologies such as network software architectures, middleware, and protocol stacks. The software-based network technologies become the main driving force behind the paradigm shift in this domain and have invented many new network technologies such as grid computing, cloud computing, fog computing, edge computing, software-defined networks, and content-centric networks. On the other hand, a lot of efforts have been made in cellular network technologies to improve the user experience, and as a consequence, emerging cellular network technologies like LTE, VoLTE, and 5G have been invented. Due to its demand and importance in the present and future scenarios, numerous efforts have been done in the networking domain by the researchers, a lot of works are still ongoing, and many more possibilities have still to be explored. Therefore, there is a need to keep track of advancements related to the network technologies and further investigate several research challenges to overcome for the ease of users. With this goal, the book provides the most recent and prominent research works that have been done related to the network technologies.

TARGET AUDIENCE

This book will be beneficial for academicians, researchers, developers, engineers, and practitioners working in or interested in the research trends and applications of network technologies. This book is expected to serve as a reference book for developers and engineers working in the computer and cellular networks domain and for a graduate/postgraduate course in computer science and engineering/information technology/electronics and communication engineering.

MATLAB® is a registered trademark of The MathWorks, Inc. For product information, please contact:
The MathWorks, Inc.
3 Apple Hill Drive
Natick, MA 01760-2098 USA
Tel: 508-647-7000
Fax: 508-647-7001
E-mail: info@mathworks.com
Web: www.mathworks.com

Acknowledgments

We are extremely thankful to the authors of the 19 chapters of this book, who have worked very hard to bring this unique resource forward for helping the students, researchers, and community practitioners. We feel that it is contextual to mention that as the individual chapters of this book are written by different authors, the responsibility of the contents of each of the chapters lies with the concerned authors.

We like to thank Randi Cohen, Publisher—Computer Science and IT, and Gabriella Williams, Editor, who worked with us on the project from the beginning, for his professionalism. We also thank all the team members of the publisher who tirelessly worked with us and helped us in the publication process.

This book is a part of the research work funded by "Seed Grant to Faculty Members under IoE Scheme (under Dev. Scheme No. 6031)" awarded to Anshul Verma at Banaras Hindu University, Varanasi, India.

Editors

Dr. Anshul Verma received M.Tech. and Ph.D. degrees in Computer Science and Engineering from ABV-Indian Institute of Information Technology and Management Gwalior, India. He has done post-doctorate from the Indian Institute of Technology Kharagpur, India. Currently, he is serving as an assistant professor in the Department of Computer Science, Institute of Science, Banaras Hindu University Varanasi, India. He has also served as a faculty member in the Computer Science and Engineering Department at Motilal Nehru National Institute of Technology (MNNIT) Allahabad and National Institute of Technology (NIT) Jamshedpur, India. His research interests include IoT, mobile ad hoc networks, distributed systems, formal verification, and mobile computing. He is serving as editor of the *Journal of Scientific Research of the Banaras Hindu University.*

Dr. Pradeepika Verma received her Ph.D. degree in Computer Science and Engineering from the Indian Institute of Technology (ISM) Dhanbad, India. She has received M.Tech in Computer Science and Engineering from Banasthali University, Rajasthan, India. Currently, she is working as a faculty fellow in Technical Innovation Hub at Indian Institute of Technology, Patna, India. She has worked as a post-doctoral fellow in Department of Computer Science and Engineering at Indian Institute of Technology (BHU), Varanasi, India. She has also worked as an assistant professor in the Department of Computer Science and Engineering at Pranveer Singh Institute of Technology, Kanpur, India, and as a faculty member in the Department of Computer Application at the Institute of Engineering and Technology, Lucknow, India. Her current research interests include IoT, natural language processing, optimization approaches, and information retrieval.

Dr. Kiran Kumar Pattanaik holds Bachelor degree in Electrical and Electronics Engineering, followed by Master and Doctorate in Computer Science and Engineering. He is presently working as associate professor at ABV-Indian Institute of Information Technology and Management, Gwalior, India. He is enthusiastic about exploring various engineering application domains involving distributed and mobile computing, wireless sensor network protocols, IoT, and edge computing. His wireless sensor network laboratory is equipped with the necessary computing facilities (simulation and hardware) and is accessible round the clock for learning. The competitive ambiance of the laboratory is instrumental for a number of publicly funded research projects and high impact factor research publications. Dr. Pattanaik (Senior Member of IEEE) is a reviewer for several leading journals in the areas of communication and networking.

Dr. Lalit Garg is a senior lecturer in Computer Information Systems at the University of Malta, Msida, Malta. He is also an Honorary Lecturer at the University of Liverpool, UK. He has also worked as a researcher at the Nanyang Technological University, Singapore, and the University of Ulster, UK. He is skilled in solving complex problems using machine learning and data analytics, especially from medicine and healthcare domain. He has published over 160 technical papers in refereed high impact journals, conferences, and books and has more than 1600 citation counts to publications, and some of his articles were awarded best paper awards. He has delivered more than 40 keynote speeches in different countries, organized/chaired/co-chaired 16 international conferences, and contributed as a technical committee member or a reviewer of several high impact journals and reputed conferences. He is awarded research studentship in Healthcare Modelling to carry out his Ph.D. research studies in the Faculty of Computing and Engineering at the University of Ulster, UK. He completed his post-graduate work in Information Technology from ABV-IIITM, Gwalior, India, and received his first degree in Electronics and Communication Engineering from Barkatullah University, Bhopal, India. His research interests are business intelligence, machine learning, data science, deep learning, cloud computing, mobile computing, the Internet of Things (IoT), information systems, management science and their applications, mainly in healthcare and medical domains. He participates in many EU, and local funded projects, including a one million euro Erasmus+ Capacity-Building project in Higher Education (CBHE) titled Training for Medical education via innovative eTechnology (MediTec), and Malta Council of science and technology's Space Research Funds. The University of Malta has awarded him the 2021-22 Research Excellence Award for exploring Novel Intelligent Computing Methods for healthcare requirements forecasting, allocation and management (NICE-Healthcare).

Contributors

Sasmita Acharya
Department of Computer Application
Veer Surendra Sai University of
    Technology
Burla, India

Anand R.
Department of Computer Science and
    Engineering
BMS Institute of Technology and
    Management
Bengaluru, India

Anil G. N.
Department of Computer Science and
    Engineering
BMS Institute of Technology and
    Management
Bengaluru, India

Prajwal Bhardwaj
Department of Information Systems
University of Texas
Austin, Texas

Ajay Kumar Bharti
Department of Computer Science and
    Application
Babu Banarasi Das University
Lucknow, India

Amit Bhatt
Information and Communication
    Technology (ICT)
Dhirubhai Ambani Institute of
    Information and Communication
    Technology
Gandhinagar, India

Haritha Bose
Amrita Mind Brain Center
and
School of Biotechnology
Amrita Vishwa Vidyapeetham
Amritapuri Campus, Kollam, India

Stevina Correia
Department of Information Technology
Dwarkadas J. Sanghvi College of
    Engineering
Mumbai, India

Saloni Dagli
Department of Information Technology
Dwarkadas J. Sanghvi College of
    Engineering
Mumbai, India

Kashvi Dedhia
Department of Information Technology
Dwarkadas J. Sanghvi College of
    Engineering
Mumbai, India

Deepak D.
Department of Computer Science and
    Engineering
Canara Engineering College
Bantwal, India

Devidas B
Department of Information Science and
    Engineering
NMAM Institute of Technology, NITTE
    (Deemed to be University)
Karkala, India

Shyam Diwakar
Amrita Mind Brain Center
and
School of Biotechnology
Amrita Vishwa Vidyapeetham
Amritapuri Campus, Kollam, India

Riddham Gadia
Department of Information Technology
Dwarkadas J. Sanghvi College of
   Engineering
Mumbai, India

M. Geetha
Department of Computer Science and
   Engineering
Kongu Engineering College
Erode, India

Sree Bhanu Gummadi
Department of Computer Science and
   Engineering
VR Siddhartha Engineering College
Vijayawada, India

Jeyashree G.
Department of Information Technology
Thiagarajar College of Engineering
Madurai, India

Abhijit Joshi
Department of Information Technology
Dwarkadas J. Sanghvi College of
   Engineering
Mumbai, India

Shivam Kejriwal
Department of Information Technology
Dwarkadas J. Sanghvi College of
   Engineering
Mumbai, India

Kaustubh Lohani
Department of Computer Science,
Stevens Institute of Technology
Hoboken, New Jersey, United States

Dharmendra Prasad Mahato
Department of Computer Science &
   Engineering
National Institute of Technology
Hamirpur, India
and
Department of Information Technology
Ton Duc Thang University
Ho Chi Minh City, Vietnam

Aditya Manchanda
Department of Computer Science and
   Engineering
Dr. B.R. Ambedkar National Institute of
   Technology
Jalandhar, India

Amit Mankodi
Information and Communication
   Technology (ICT)
Dhirubhai Ambani Institute of
   Information and Communication
   Technology
Gandhinagar, India

Muneshwara M. S.
Department of Computer Science and
   Engineering
BMS Institute of Technology and
   Management
Bengaluru, India

Muskan
Department of Computer Science &
   Engineering
National Institute of Technology
Hamirpur, India

S. K. Nivetha
Department of Computer Science and
   Engineering
Kongu Engineering College
Erode, India

Padmavathi S.
Department of Information Technology
Thiagarajar College of Engineering
Madurai, India

Vasudeva Pai
Department of Information Science and
    Engineering
NMAM Institute of Technology, NITTE
    (Deemed to be University)
Karkala, India

Ramesh Kumar Panneerselvam
Department of Computer Science and
    Engineering
VR Siddhartha Engineering College
Vijayawada, India

Vikas Kumar Patel
Department of Computer Science
Banaras Hindu University
Varanasi, India

Van Huy Pham
Department of Information Technology
Ton Duc Thang University
Ho Chi Minh City, Vietnam

Dheeraj Pisharody
Amrita Mind Brain Center
and
School of Biotechnology
Amrita Vishwa Vidyapeetham
Amritapuri Campus, Kollam, India

Mohammed Rafi
Department of Computer Science and
    Engineering
VR Siddhartha Engineering College
Vijayawada, India

Dijith Rajan
Amrita Mind Brain Center
and
School of Biotechnology
Amrita Vishwa Vidyapeetham
Amritapuri Campus, Kollam, India

Rashmi Ranjita
Department of Computer Application
Veer Surendra Sai University of
    Technology
Burla, India

Abhishek S. Rao
Department of Information Science and
    Engineering
NMAM Institute of Technology, NITTE
    (Deemed to be University)
Karkala, India

Shailendra Kumar Rawat
AIIT
Amity University
Lucknow Campus, Uttar Pradesh, India

Hemalatha Sasidharakurup
Amrita Mind Brain Center
and
School of Biotechnology
Amrita Vishwa Vidyapeetham
Amritapuri Campus, Kollam, India**N.**
    Senthilkumaran
Department of Computer Applications
Vellalar College for Women
Erode, India

Rushank Shah
Department of Information Technology
Dwarkadas J. Sanghvi College of
    Engineering
Mumbai, India

Chilka Sharma
School of Earth Sciences
Banasthali Vidyapith
Rajasthan, India

Rashmi Sharma
Shri Vaishnav Institute of Computer
    Applications
Shri Vaishnav Vidyapeeth
    Vishwavidyalaya
Indore, India

Uma Sharma
Department of Computer Science
Banasthali Vidyapith
Rajasthan, India

Govind Sreekar Shenoy
Department of Information Science
 Engineering
Nitte Meenakshi Institute of Technology
Bangalore, India

Nagesh Shenoy
Department of Computer Science and
 Engineering
Canara Engineering College
Bantwal, India

Karthik Sainadh Siddabathula
Department of Computer Science and
 Engineering
VR Siddhartha Engineering College
Vijayawada, India

Akhilendra Pratap Singh
School of Engineering and Technology
Maharishi University of Information
 Technology
Lucknow, India

Amrit Pal Singh
Department of Computer Science and
 Engineering
Dr. B.R. Ambedkar National Institute of
 Technology
Jalandhar, India

Mohini Singh
Department of Computer Science
Banaras Hindu University
Varanasi, India

S. K. Singh
AIIT
Amity University
Lucknow Campus, Uttar Pradesh, India

B. Srinivasa Rao
Department of Computer Science and
 Engineering
Gokaraju Rangaraju Institute of
 Engineering and Technology
Bachupally, India

R. C. Suganthe
Department of Computer Science and
 Engineering
Kongu Engineering College
Erode, India

Het Suthar
Information and Communication
 Technology (ICT)
Dhirubhai Ambani Institute of
 Information and Communication
 Technology
Gandhinagar, India

Swetha M. S.
Department of Information Science and
 Engineering
BMS Institute of Technology and
 Management
Bengaluru, India

Vineela Vasana
Department of Computer Science and
 Engineering
VR Siddhartha Engineering College
Vijayawada, India

Jayavardhan Vejendla
Department of Computer Science and
 Engineering
VR Siddhartha Engineering College
Vijayawada, India

Anshul Verma
Department of Computer Science
Banaras Hindu University
Varanasi, India

Harsha Verma
Department of Computer Science and
    Engineering
Dr. B.R. Ambedkar National Institute of
    Technology
Jalandhar, India

Pradeepika Verma
Technology Innovation Hub
Indian Institute of Technology
Patna, India

1 Energy and Delay Balance Ensemble Scheduling Algorithm for Wireless Sensor Networks

B. Srinivasa Rao
Gokaraju Rangaraju Institute of Engineering and Technology

CONTENTS

1.1   INTRODUCTION

The most important aspect of the wireless sensor networks (WSNs) is the necessity of long-term involvement and working of sensor node batteries without charging while monitoring the critical events that affect the efficiency of the WSN. Hence, intelligent techniques are required for effective conservation of the energy of the power sources. Based upon the energy waste in WSN, various types of methods like data-reduction, control-reduction, energy-efficient routing, duty-cycling, and topology-control have been reported in the literature for both general WSNs and mobile-sink-based WSNs [1–8]. Analytically, all these techniques have their advantages and disadvantages at a wide range [2]. Obviously, it has been observed that it is essential to design energy-efficient scheduling algorithms to enhance the lifetime of the power source and in turn of the sensor nodes [3]. In that context, sleep scheduling algorithms significantly reduced WSNs' energy consumption and time delay [4]. In general, sleep scheduling algorithms are used in the form of synchronous,

DOI: 10.1201/9781003320333-1

semi-synchronous, and asynchronous mechanisms. In a synchronous scheduling mechanism, all sleeping nodes wake up for communication and require additional control traffic. In a semi-synchronous mechanism, the nodes form into clusters, and the sleeping and wake-up occur within the clusters. But all these clusters are not in synchronization. In asynchronous sleep scheduling, each node contains its scheduling as per requirement [5]. Most of the sleep scheduling algorithms fall into any one of these mechanisms. Recently, optimization techniques and heuristic approaches have also become very popular for efficient sleep scheduling mechanisms [9–11]. The main aim of the present research is to propose and implement a novel efficient energy and delay balance ensemble scheduling algorithm for WSNs using two popular search techniques, breadth first search (BFS) and color-connected dominated set (CCDS), for reducing energy consumption and delay when a message is broadcasted in WSN. The novelty in the present research problem is an ensemble of BFS algorithm and CCDS algorithm, which earlier researchers did not consider. BFS is implemented to find the minimum distance path from a sensor node and reduce the delay in transmitting the message. CCDS is used to transmit messages to all nodes without collision and hence minimize the energy consumption. In the present paper, the concepts of BFS and CCDS scheduling are given briefly in Section 1.2. The related work has been presented in Section 1.3. The proposed energy and delay balance ensemble scheduling algorithm for WSNs has been described in Section 1.4. Methodology and experimental parts are given in Sections 1.5 and 1.6, respectively. Section 1.7 deals with results and discussion parts. Finally, the work is concluded in Section 1.8.

1.2   SCHEDULING USING BFS AND CCDS

This section describes the working of BFS and CCDS scheduling algorithms.

1.2.1   BFS SCHEDULING

The BFS scheduling is designed from the inspiration of the BFS algorithm of the graph theory. The procedure begins with visiting the sensor node and all of its neighbor sensors. Then, in the subsequent step, the nearest neighboring sensors of the nodes are visited, and the same procedure is continued in the next subsequent steps. The algorithm visits all the adjacent sensor nodes of all the sensor nodes in the network and ensures that each sensor is visited exactly once [12]. By implementing the BFS algorithm, a BFS tree is constructed for the uplink path in the following steps.

Step 1. A sensor node is selected as the central node.
Step 2. Categorization of all the sensor nodes into node levels L1, L2, L3...etc.
Step 3. Each level is depicted by different colors.
Step 4. In this process, the neighboring sensors of each sensor are computed, as shown in Table 1.1.

From Table 1.1, BFS tree is constructed for each node as shown in Figure 1.1. Then, a routing table is built for each node in the WSN, as shown in Table 1.2. Now, each route in the BFS tree routing table defines the uplink paths for that particular sensor node.

TABLE 1.1
Node Representation

S. No	Source Node	Neighbor Nodes
1	1	2,3,4
2	2	1,6,5
3	3	1,4,5,8
4	4	3,1,7
5	8	3,7

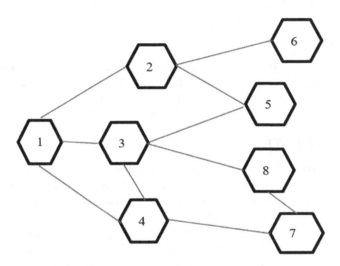

FIGURE 1.1 BFS tree structure.

TABLE 1.2
Routing Table and Uplink Paths

S. No	Source Node	Uplink Path
1	1	1->1
2	2	2->1
3	3	3->1
4	4	4->1
5	5	5->2->1 or 5->3->1
6	6	6->2->1
7	7	7->4->1
8	8	8->3->1

1.2.2   Color-Connected Dominant Set (CCDS) Scheduling

The CCDS scheduling is used to construct a downlink path in WSN when a critical event has occurred. The scheduling design is from the inspiration of the CCDS concept proposed for WSN that implements self-regulation among the nodes [1]. In WSNs, the downlink path construction is different with respect to uplink paths as the communication will be between a long-distance node to the central node and possibility of multiple paths and to select an optimal one. This process requires internal self-regulation among the sensor nodes. In the present CCDS scheduling, different algorithms are combined to construct a connected dominated set (CDS) and serve as a backbone to the WSN. The main objective of the backbone is the reduction of communication overhead, enhancement of bandwidth efficiency, minimization of overall energy consumption, and improvement of network effective lifetime in a WSN [1].

1.2.3   Construction

For the construction of CCDS, we have followed the earlier methods proposed in [13,14]. The construction process involves (i) maximum independent set (MIS) in G, (ii) CDS, and (iii) internal model control (IMC) algorithm [15].

1.3   RELATED WORK

Different methods of sleep scheduling for extension of the lifetime of the WSN have been recently reviewed [1–5, and 9–11]. In general, most of the proposed sleep scheduling mechanisms involve the components like target prediction, reduction of awakened sensors, and control active time of the sensor. The energy-efficient Time Division Multiplexing Algorithm (TDMA) sleep scheduling algorithm has the advantage of maximization of the lifetime of WSN, but this mechanism has disadvantages like delay, data overlap, and reduction in channel utilization [16]. The balanced energy scheduling used WSN sensor redundancy to increase the lifetime and network load balance to improve the efficiency [17], but it has the disadvantage of long-distance communication in the WSN. An evolutionary multiobjective sleep scheduling scheme for differentiated coverage in WSNs algorithm reduces energy utilization and communication delay but has the issues of communication delay and latency [18]. Recently, a new energy-efficient reference node selection (EERS) algorithm has been proposed [19] for time synchronization in industrial WSNs. EERS significantly increased the large savings in energy consumption but was applicable to many nodes at the industry level. A multilevel sleep scheduling algorithm was developed, adopting the clustering concept for WSNs [20]. Even though the model increases the network lifetime, asynchronization among the clusters is an issue to consider. An efficient sleep scheduling mechanism was developed for WSN using similarity measures [21]. This model reduces the energy consumption by scheduling the nodes into active or sleeping modes. But this mechanism is not effective in the case of sparse distribution of sensor nodes. A heuristic-based delay tolerance and energy saving model was developed for WSN [22], which gave a better performance to other models but is confined to only a mobile base station scenario. A sensor node scheduling algorithm was proposed for heterogeneous WSN [23] to improve network lifetime and regional coverage rate. However, this model is more suitable for only static nodes WSN and may not be effective for mobile nodes. Recently, Mhatre and Khot [24] proposed an energy saving opportunistic routing sleep scheduling algorithm that effectively reduces energy dissipation but is confined to only one-dimensional topology networks. Recently, Sinde et al. [25] proposed energy-efficient scheduling using deep reinforcement learning that increases the lifetime and reduces the network delay and has shown better performance than previous models. But the main issue with the model is with the complexity of deep reinforcement learning. Ant colony optimization algorithm was used for energy optimization of WSN with better energy efficiency [9]. Another scheduling algorithm was proposed by Manikandan [26] using game theory and wake-up approach for energy efficiency in WSN. The main disadvantage of this model is many approximations that make the model unrealistic. Recently, a metric routing protocol was proposed for the evolution and stability of Internet of things networks [10]. A heuristic approach-based ant colony optimization multipath routing algorithm was proposed for virtual ad hoc networks to optimize the relay bus and route selection issues [11]. The above two models give an idea for considering some new approaches for efficient scheduling mechanisms for WSN. Also, minimizing energy consumption is good security-providing aspect in WSN [27]. From the above discussion, we infer that the above-mentioned scheduling algorithms are good at some points and have drawbacks in other aspects and stressing further research in designing new and novel efficient algorithms to enhance the lifetime of the sensors. In all the above algorithms and models, no appropriate attempt has been made to balance energy consumption and delay time simultaneously. The main advantage of BFS in WSN is that during its traversal of the tree in the level-by-level manner, it classifies tree edges and cross edges, and its time complexity is O(N). This property is very important for efficient routing in WSNs [12]. Also, BFS is implemented to find the minimum distance path from a sensor node and reduce the delay in transmitting the message. On the other hand, the CDS-based routing is one kind of hierarchical method that has received more attention in reducing routing overhead. CCDS is used to transmit messages to all nodes without collision and hence minimize the energy consumption [1]. The ensemble of these two techniques balances both energy consumption and delay time in the WSN. In the present research work, we propose an efficient sleep scheduling procedure for WSN using two popular search techniques, BFS and CCDS, for reducing energy consumption and delay when the message is broadcasted in WSN. In the next section, the proposed model has been presented.

1.4 ENERGY AND DELAY BALANCE ENSEMBLE SCHEDULING ALGORITHM

In the present paper, we propose an efficient sleep scheduling procedure for WSNs using two popular search techniques, BFS and CCDS, for reducing energy consumption and delay when the message is broadcasted in WSN. The proposed algorithm

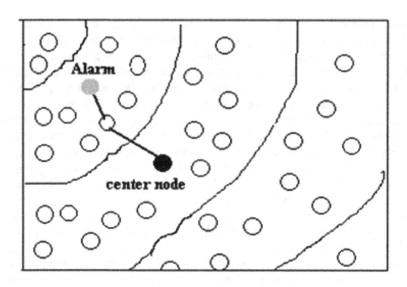

FIGURE 1.2   Construction of uplink path.

is considered in two phases: (a) uplink phase and (b) downlink phase for scheduling. The uplink phase is scheduled by BFS scheduling with the inspiration of graph theory's BFS algorithms. CCDS schedules the downlink phase-inspired scheduling. Finally, the combination of BFS scheduling and CCDS scheduling forms the proposed novel efficient sleep scheduling algorithm to reduce energy consumption and delay when message is broadcasted in WSN. The proposed new scheduling is briefly described as follows. A model WSN that has been deployed for the detection of any critical event is as shown in Figure 1.2. It consists of a central node (black-shaded) also known as center node that has the capability of communication with all the network nodes. In the case of detecting a critical or disaster event by any node of the network (denoted by gray-shaded node), the gray node sends an alarm to the central node and thus constructs the uplink path. For the construction of the uplink path, BFS scheduling is implemented. In constructing this uplink path, the shortest path from any node to the central node is computed by BFS scheduling.

Now, the central node transmits the received alert from the gray node during the uplink phase to all the other sensors in the network. For the construction of the downlink path, the CCDS scheduling is implemented. For the construction of the downlink path, CCDS is constructed using the IMC algorithm [1,15]. The IMC algorithm is self-regulating process and characterizes the downlink path while transmitting the alert from the central node to all other sensor nodes of WSN, as shown in Figure 1.3.

1.5   METHODOLOGY

This section discusses the methodology for both uplink and downlink phases of the present scheduling algorithm. Initially, deployment of nodes is performed. In order to have communication in the WSN, route discovery is done using route table entries

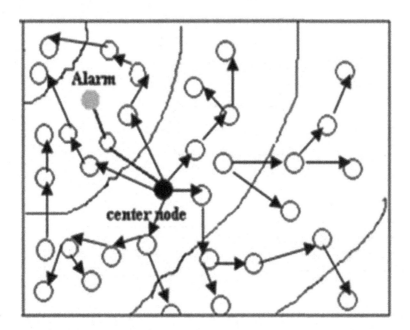

FIGURE 1.3 Construction of downlink path.

of all the sensor nodes in WSN. A node initiates route discovery by sending a request to its neighboring first hop nodes to know whether they are located in its path and waits for a route reply from the first hop neighbors. Based on the first hop nodes' reply messages, the broadcasting node updates its routing table entry destination ID. The sequence number and battery status are updated for the latest information about the nodes' fresh routes and energy levels. After identifying the neighbor nodes, the BFS algorithm is implemented to construct the BFS tree that divides the nodes into different levels. Using these levels, a CCDS is constructed. Also, MIS is constructed for independent nodes of each level. For a comparative study, both scheduling algorithms will be inputted with the same set of nodes. The methodology diagram for the proposed work is as shown in Figure 1.4.

The pseudocode for phases of BFS scheduling and CCDS construction has been presented below. The general notations are followed in the pseudocodes.

Pseudocode 1. BFS Scheduling Procedure for WSN denoted by V [12] (For notations, refer to [12])

```
begin
for each node n ε V do
Distance[n]= infinity, Predicate[n]=-1;
Color[n]=White;
Distance[s]=0, Color[s]=Gray;
Q=Empty, Queue Enqueue(Q, s);
while(Q is not empty) do
u=Head(Q);
for each neighbor of n of u do
```

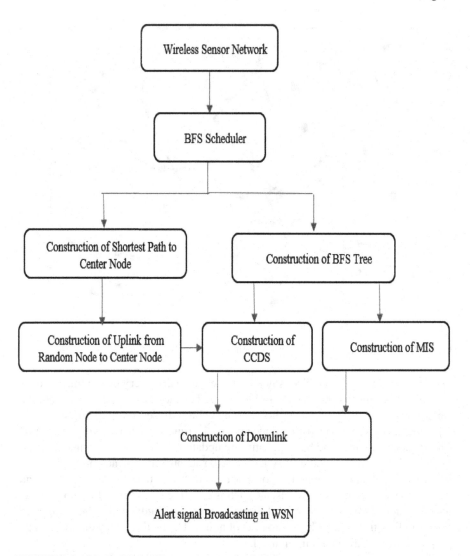

FIGURE 1.4 Methodology diagram for proposed work.

```
if(Color[n] is White) then
Distance[n]=Distance[u] + 1;
Predicate[n]=u;
Color[n]=Gray;
Enqueue(Q, n);
Dequeue(Q);
Color[u]=Black;
end;
Pseudocode 2: Construction of Maximum Independent Set [13]
(For notations, refer to [13])
Function MIS(W)
```

```
begin
if (!connected(W)) then
begin
X=SCC(W);
if(|X|<=2) P=1 else P=MIS(X);
return (MIS(W-X) + P);
end;
if (|W|<=1) then return(|W|);
Select Y, Z of W such that
(i)d(Y, W) is minimal and
(ii) (Y, Z) is an Edge of W and d(Z, W) is maximal for all
neighbors with degree d(Y, W);
if(d(Y, W)=1) then return 1+ MIS(W- M(Y);
if (d(Y, W)=2 then
begin
Z':=M(Y)-Z;
if (Edge(Z, Z')) then return (1+ MIS(W-M(Y));
return Maximum(2+MIS(W-M(Z)-M(Z')), 1 + MIS² (W-M(Y)), M²(Y));
end;
if (d(Y, W)=3) then return Maximum(MIS2 (W-Y, M(Y)), 1+
MIS(W- M(Y)));
if Y dominates Z then return MIS(W-Z);
return Maximum(MIS(W-Z), 1 + MIS(W-M(Z)))
end;
```

1.6 EXPERIMENT

The software requirements are (a) Backend: Python 3 and (b) Frontend: HTML, CSS, and Bootstrap 4. The hardware requirements are RAM—advised to have >32GB, Graphic Card, Nvidia GTX 1071, Processor—Intel Core i7-8750H, Storage—512GB SSD. Simulator: NetSim. Typical sequence of steps to do experiment: (1). Network Set up: Drag and drop devices, and connect them using wired or wireless links. (2). Configure Properties: Configure device, protocol, or link properties by right-clicking on the device or link and modifying parameters in the properties window. (3). Model Traffic: Click on the Application icon present on the ribbon and set traffic flows. (4). Enable Trace/Plots : Click on packet trace, event trace, and Plots to enable. Packet trace logs packet flow, event trace logs each event, and the Plots button enables charting of various throughputs over time. (5). Save/Save As/Open/Edit: Click on File à Save / File à Save As to save the experiments in the current workspace. Saved experiments can then be opened from NetSim home screen to run the simulation or to modify the parameters and again run the simulation. (6). View Animation/View Results: Visualize through the animator to understand working and to analyze results and draw inferences (Figure 1.5).

1.7 RESULTS AND DISCUSSION

A novel efficient sleep scheduling algorithm for WSN was proposed and experimented on network simulator NetSim. The NetSim has been set up and connected to the designed model WSN. The system was configured for the routing protocols

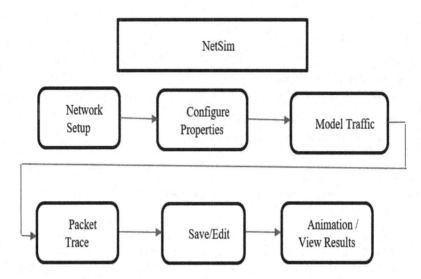

FIGURE 1.5 NetSim experiment for the present ensemble model.

presented in the model. With a set of model inputs, simulation has been done and output has been tabulated. The information of the source node and its neighbors has been inputted and given in Table 1.3. The implemented results of the BFS scheduling algorithm are presented in Table 1.4. Similarly, the experimental results of the CCDS scheduling algorithm are given in Table 1.5. Table 1.6 shows the paths from any sensor node to the central node in WSN. From Table 1.4, it can be observed that during BFS scheduling, some parent nodes have children, and others do not have as they are leaf nodes. It is well understood that no energy is released from leaf nodes.

Similarly, from Table 1.5, it is understood that during CCDS scheduling, the parent nodes not having children would not be involved in message transmission. Comparing Tables 1.4 and 1.5, it can be understood that the child node number of the parent nodes of BFS scheduling and CCDS scheduling is not the same.

TABLE 1.3
Input Neighbors List

S. No	Nodes	Adjacent List
1	P1	P2, P3, P4, P5, P6, P7
2	P2	P8, P9, P11, P3
3	P6	P7, P18, P19, P17
4	P8	P9, P20
5	P7	P8, P19, P20
6	P4	P15, P14, P13
7	P3	P12, P4, P13
8	P12	P13
9	P13	P14

TABLE 1.4
Output of BFS

S. No	Parent Node	Child Node
1	P1	P6, P5, P4, P7, P2, P3
2	P6	P19, P17, P18
3	P7	P20, P8
4	P8	-
5	P9	P22, P21, P23
6	P13	-
7	P4	P13, P14
8	P14	-
9	P17	P16
10	P5	P15
11	P15	-
12	P18	-
13	P3	P12
14	P12	-
15	P2	P11, P9

TABLE 1.5
Output of CCDS

S. No	Parent Node	Child Node
1	P1	P6, P5, P4, P7, P2, P3
2	P6	P17
3	P7	P20, P19
4	P8	-
5	P9	P21, P22, P23
6	P13	P12, P14
7	P4	P13, P15
8	P14	-
9	P17	P16
10	P5	-
11	P15	-
12	P18	-
13	P3	-
14	P12	-
15	P2	P11, P9

Also, childless nodes are more for CCDS. Therefore, finally, it could be understood that the childless parents of BFS do not involve in the dissipation of energy as they are leaf nodes, and childless nodes of CCDS do not involve in transmission message and turn conserve energy, which is the main objective of the present research work.

TABLE 1.6

Paths Representation

Node to Center Node	BFS Path to Center Node	CCDS Path to Center
P15-P1	P15-P5-P1	P15-P4-P1
P13-P1	P13-P4-P1	P13-P4-P1
P12-P1	**P12-P3-P1**	**P12-P13-P4-P1**
P14-P1	**P14-P4-P1**	**P14-P13-P4-P1**
P8-P1	**P8-P7-P1**	**P8-P20-P7-P1**
P9-P1	P9-P2-P1	P9-P2-P1
P17-P1	P17-P6-P1	P17-P6-P1

Table 1.6 shows the paths from any sensor node to the central node in WSN. The bold-lettered nodes in the table reveal the difference in the number of hops. It can be observed that the sensor nodes P12, P14, and P8 have a smaller number of hops in path to the central node N1 for BFS scheduling in comparison with CCDS scheduling. At the same time, the nodes P3, P4, P5, P6, and P7 are at one-hop distance from P1. Other nodes such as P2, P3, P5, P6, and P7 are one hop away from the center node. It is obvious that the hop count varies with the levels of the nodes in the WSN.

1.8   CONCLUSION

In the present paper, the proposed energy and delay balance ensemble scheduling algorithm has been successfully designed and implemented. From the experimental results, it can be concluded that to have balance and better energy saving and fastest transmission of a message in WSN, the present algorithm has successfully balanced both the BFS and CCDS. Further research is required to improve the model. In the future, we extend this model for both homogeneous and heterogeneous networks of sufficiently large size and it would be compared with other models numerically and analytically.

Acknowledgments: The author is thankful to the management of GRIET for their encouragement.

REFERENCES

1. Liu, Z., Wang, B., and Guo, L., "A survey on connected dominating set construction algorithm for wireless sensor networks", *Information Technology Journal*, 9(6), 1081–1092 (2010).
2. Soua, R. and Minet, P., "A survey on energy efficient techniques in wireless sensor networks", *2011 4th Joint IFIP Wireless and Mobile Networking Conference (WMNC 2011)*, 2011, pp. 1–9. doi: 10.1109/WMNC.2011.6097244.
3. Pagar, A.R. and Mehetre, D.C., "A survey of energy efficient sleep scheduling in WSN", Semanticscholar.org, Corpus ID 212548604 (2015).
4. Karthihadevi, M. and Pavalarajan, S., "Sleep scheduling strategies in wireless sensor network", *Advances in Natural and Applied Sciences*, 11(7), 635–641 (2017).

5. Zhang, Z., Shu, L., Zhu, C., and Mukherjee, M., "A short review on sleep scheduling mecha-nism in wireless sensor networks", *International Conference on Heterogeneous Networking for Quality, Reliability, Security and Robustness*, 2017, pp. 66-70, Springer, Cham.
6. Jain, S., Pattanaik, K.K., Verma, R.K., Bharti, S., and Shukla, A., "Delay-aware green routing for mobile-sink-based wireless sensor networks", *IEEE Internet of Things Journal*, 8(6), pp. 4882–4892 (2021). doi: 10.1109/JIOT.2020.3030120.
7. Jain, S., Pattanaik, K.K., Verma, R.K. et al., "EDVWDD: Event-driven virtual wheel-based data dissemination for mobile sink-enabled wireless sensor networks", *Journal of Supercomputing*, 77, 11432–11457 (2021). doi: 10.1007/s11227-021-03714-7.
8. Jain, S.K., Venkatadari, M., Shrivastava, N. et al., "NHCDRA: A non-uniform hierar-chical clustering with dynamic route adjustment for mobile sink based heterogeneous wireless sensor networks", *Wireless Networks*, 27, 2451–2467 (2021). doi: 10.1007/s11276-021-02585-3.
9. Chen, J.I.Z. and Lai, K., "Machine learning based energy management at internet of things network nodes", *Journal of Trends in Computer Science and Smart Technology*, 2(3), 127–133 (2020).
10. Smys, S. and Vijesh Joe, C., "Metric routing protocol for detecting untrustworthy nodes for packet transmission", *Journal of Information Technology*, 3(2), 67 (2021).
11. Dhaya, R. and Kanthavel, R., "Bus-based VANET using ACO multipath routing algo-rithm", *Journal of Trends in Computer Science and Smart Technology (TCSST)*, 3(1), 40 (2021).
12. Akram, V. K. and Dagdeviren, O. "Breadth-first search-based single-phase algorithms for bridge detection in wireless sensor networks", *Sensors (Basel, Switzerland)*, 13(7), 8786–8813 (2013). doi: 10.3390/s130708786.
13. Robson, J.M., "Algorithms for maximum independent sets", *Journal of Algorithms*, 7, 425–440 (1986).
14. Peng, G., Tao, J., Qian, Z., and Kui, Z., "Sleep scheduling for critical event monitoring in wireless sensor networks", *IEEE Transactions on Parallel and Distributed Systems*, 23(2) (2012).
15. Rivera, D., Morari, M., and Skogestad, S. "Internal model control - PID controller design", *Industrial & Engineering Chemistry Process Design and Development*, 25, 252–265 (1986).
16. Laxman, P. and Rajeev, P., "Comparative analysis of TDMA scheduling algorithms in wireless sensor networks", https://www.semanticscholar.org/, Corpus ID: 61778529 (2014).
17. Feng, J. and Zhao, H., "Energy balanced multisensory sensory scheduling for target tracking in WSN", *Sensors (Basel)*, 18(10), 3585 (2018).
18. Soumyadip, S., Swagatam, D., Nasir, M., Athanasios, V., and Witold, P., "An evolu-tionary multiobjective sleep-scheduling scheme for differentiated coverage in wire-less sensor networks", *IEEE Transactions on Systems, Man, and Cybernetics, Part C (Applications and Reviews)*, 42(6), 1093–1102 (2012).
19. Elsharief, M., El-Gawad, M.A.A., Ko, H., and Pack, S., EERS: "Energy-efficient refer-ence node selection algorithm for synchronization in industrial wireless sensor net-works", *Sensors*, 20, 4095 (2020). doi: 10.3390/s20154095.
20. Hassan, S., Nisar, M.S., and Jiang, H., "Multilevel sleep scheduling for heterogeneous wireless sensor networks", *Computer Science, Technology and Application*, 227 (2016).
21. Wan, R., Xiong, N., and Loc, N.T., "An energy efficient sleep scheduling mechanism with similarity measure for wireless sensor networks", *Human-Centric Computing and Information Sciences*, 8, 18 (2018).
22. Jerew, O. and Bassan, N., "Delay tolerance and energy saving in WSN in mobile base station", *Wireless Communication and Mobile Computing*, 2019, 3929876, 12 pages (2019). https://doi.org/10.1155/2019/3929876

23. Wang, Z., Chen, Y., Liu, B., Yang, H., Su, Z., and Zhu, Y., "A sensor node scheduling algorithm for heterogeneous wireless sensor networks", *International Journal of Distributed Sensor Networks*, 15, 1 (2019).

24. Mhatre, K.P. and Khot, U.P., "Energy efficient opportunistic routing with sleep scheduling in wireless sensor networks", *Wireless Personal Communications*, 112, 1243 (2020).

25. Sinde, R., Begum, F., Njau, K., and Kaijage, S., "Refining network life time of wireless sensor networks using energy efficient clustering and DRL based sleep scheduling", *Sensors*, 20(5), 1540 (2020).

26. Manikandan, K.B., "Game theory and wake up approach scheduling in WSN for energy efficiency", *Turkish Journal of Computer and Mathematics Education*, 12(10), 2922 (2021).

27. Gudivada, R.B. and Hansdah, R.C., "Energy efficient secure communication in wireless sensor networks", *2018 IEEE 32nd International Conference on Advanced Information Networking and Applications (AINA)*, 2018, pp. 311–319, doi: 10.1109/ AINA.2018.00055.

2 An Improved RSSI-Based WSN Localization Algorithm for Coal Mine Underground Miners

Shailendra Kumar Rawat
Amity University

Akhilendra Pratap Singh
Maharishi University of Information Technology

S. K. Singh
Amity University

Ajay Kumar Bharti
Babu Banarasi Das University

CONTENTS

2.1 INTRODUCTION

The safety issue of underground miners has been a significant factor limiting the growth of the coal industry in recent decades. Coal mine accidents happen annually, which causes heavy losses of life and property. To address such issue requires the development of underground mining environmental monitoring system [1–4] and robust localization system that would minimize losses and ensure miners' safety. Recently, the use of wireless sensor networks (WSNs) has gained a lot of attention due to its benefits of self-organization, distributed activity, high reliability, and low

DOI: 10.1201/9781003320333-2

cost [5,6]. WSN location service is pivotal for the availability of critical services. The majority of the sensory data including position information is important, especially real-time subsurface data regarding the miners' locations. Therefore, to understand the exact localization, many localization algorithms and frameworks are suggested to understand different techniques in WSNs. A rather distinct application is the localization of WSNs between the open world and the underground climate of coal mines. In general, in the coal mine tunnel environment, there is a multipath effect and shadowing, so an acceptable approach needs to be chosen to consider the underground location.

Ranging-based localization and ranging-free localization are basically the two types of localization procedures [7]. In ranging-based techniques using the distances information between anchor nodes and unknown node/dumb node, one can estimate the position of unknown node. The distance information is mainly estimated by received signal strength indicator (RSSI) value. Some of the ranging-based techniques are multilateration [8], weighted centroid localization algorithm [9], ad hoc network-based localization systems [10], and difference-based hyperbolic positioning systems [11]. On the other hand, ranging-free localization strategies estimate the position of unknown node by solving optimization problems rather than using distance information [12]. Distance vector hop [8,13,14], approximation point in triangle [8,15], centroid [8], rectangular intersection [16], circular intersection [16], and hexagonal intersection [16] are some of the approaches in this category.

In harsh environment such as underground coal mine, a robust localization system is required having a significant higher localization accuracy. Since the range-based localization algorithm is having the ability to achieve higher localization accuracy compared to range free, it has been a topic of much interest among researchers in recent years. Furthermore, due to the paucity of hardware resources and inexpensive cost, the RSSI-based range technique has gotten a lot of attention. There are two key reasons that cause positioning errors in RSSI-based positioning systems. To begin with, the signal intensity of a radio transmission decreases as the propagation distance increases. In different situations, RSSI has varied amplitude of attenuation. Second, because it is affected by unfavorable environments such as multipath, diffraction, and obstacles, the RSSI value is particularly unstable. The statistical mean model, weighted model, Gaussian model, and other methods are the most common ways to generate precise RSSI values. Numerous models were simulated and found that the Gaussian model has the maximum range precision [17]. A correction technique based on Gaussian anchor nodes was developed in Ref. [18]. A deviation-based Gaussian model that can filter the variance-level interference was proposed in Ref. [19]. In Ref. [20], the RSSI value was optimized using a hybrid filtering method. The RSSI value was optimized using a Gaussian weighted technique in Ref. [21].

This research proposes a novel entropy-weighted model for optimizing RSSI values to acquire more accurate distances and employs a genetic algorithm method to provide the location of the unknown node. The paper is structured as follows: The wireless communication and RSSI range models are described in Section 2.2. The proposed RSSI enhancement and localization approach is described in Section 2.3. The simulation results are shown in Section 2.4, and conclusions are presented in Section 2.5.

2.2 WIRELESS COMMUNICATION MODEL
AND RSSI RANGING MODEL

Let A, B, and C be the beacon nodes in Figure 2.1, which are positioned at random. Consider these nodes to be wireless access points (APs) that can connect with one another via Zigbee wireless technology. When the random dumb node considered as worker comes under the communication radius of these beacon nodes, it can communicate with beacon nodes, as depicted in Figure 2.1. For example, It communicate with is three communication nodes. The dumb node measures the RSSI from each beacon node and determines the distance information of dumb node with each beacon node. After that, using distance information location of dumb node can be estimated using suitable localization algorithm. The distance from A, B, and C nodes shall not exceed the communication range to ensure the number of nodes. To obtain more precise RSSI readings, the distances can be reduced from A, B, and C. The anchor nodes can, however, be raised further if the distance is short enough. As a result, algorithm's complexity and cost will increase.

RSSI-Based Ranging Model: The strength of a radio transmission decreases as propagation distances grow. Signal strength attenuation varies depending on the environment. The free space propagation model, Hata model, log-distance distribution model, and others are popularly used wireless signal propagation models in WSNs to quantify path loss.

The propagation loss experienced by a wireless signal in an underground mining location can be predicted using a log-distance distribution model. In most real-life underground mining situations, shadowing effects are unavoidable. Therefore, a zero mean Gaussian random variable with standard deviation is included in the equation to account for the shadowing effect, which is represented as

$$P_L(d) = P_L(d_R) + 10n\log\left(\frac{d}{d_0}\right) + X_\sigma \qquad (2.1)$$

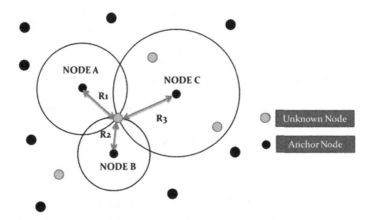

FIGURE 2.1 Wireless channel modeling.

where,

$P_L(d)$ = path loss (in dB) when transmitted signal travels a distance d,

X_σ = random variable with Gaussian distribution with zero mean value and σ standard deviation (usually takes 4–10),

n = path loss exponent (generally 2–5),

d_R = reference distance (usually 1 m), and

$P_L(d_R)$ = path loss (in dB) when transmitted signal travels distance d_R (generally 1 m).

The RSSI values of nodes at distance d can be expressed as

$$\text{RSSI} = P_T + P_G - P_L(d) \tag{2.2}$$

where

P_T, P_G = transmitted power and gain of antenna, respectively.

Similarly,

$$A = P_T + P_G - P_L(d_R) \tag{2.3}$$

where

d = RSSI at the distance d_R.

The range between unknown nodes and anchor nodes can be estimated using Equations (2.1)–(2.3):

$$d = 10^{(A-\text{RSSI}-X_\sigma)/10n} d_R \tag{2.4}$$

2.3 PROPOSED METHOD FOR RSSI IMPROVEMENT AND LOCALIZATION

When a wireless signal is propagating in a coal mine, RSSI measurements reveal a lot of randomness and volatility. The exact distance between the unknown node and anchor node cannot be determined by RSSI measurement if just formula (2.4) is used to calculate distance. The environment affects RF signals, and without taking X into account, the fluctuation range of positioning inaccuracy could be high, which is incompatible with underground positioning requirements. RSSI value improvement becomes an important aspect of the positioning algorithm to minimize positioning inaccuracy. As a result, an entropy-weighted model for RSSI correction is proposed. Then, using a genetic algorithm, determine worker position.

2.3.1 RSSI IMPROVEMENT

The steps performed to improve RSSI value are as follows:

Step 1: The unknown node records many values of RSSI with respect to a certain anchor node. A large number of RSSI values with a high probability are selected using Gaussian model [20]. The Gaussian model [20] filters the values of RSSI having a probability density of less than 0.6 (which could be due to the environment), and the residual RSSI values are ordered from lower to higher $RSSI_1 \leq RSSI_2 \leq \cdots \leq RSSI_m$, where m is the number of RSSI values left.

Step 2: Determine the weight of the RSSI values of m by calculating the entropy.

$$w_i = -p_i \log_2 p_i \tag{2.5}$$

Step 3: Calculate the correct RSSI value using Equation (2.6) expressed as

$$RSSI_C = \sum_{i=1}^{m} w_i \times RSSI_i \tag{2.6}$$

After modifying, formula (4) turns to

$$d = 10^{(A-RSSI_C)/10n} d_R \tag{2.7}$$

2.3.2 LOCALIZATION ALGORITHM

The problem of localization is modeled as optimization problem and solved with the use of genetic algorithm. Assume that the unknown node's coordinates are (x, y) and the beacon nodes' coordinates are (x_i, y_i), with $i = 1, 2, 3$. When the unknown node interacts with each beacon node, using distance information distance D_i between an unknown node and the anchor nodes, there can be three measurement equations that demonstrate

$$D_i^2 = (x - x_i)^2 + (y - y_i)^2 \quad i = 1,2,3 \tag{2.8}$$

The coordinate of unknown node can be assessed by forming a suitable objective and minimizing it using genetic algorithm

$$F_{obj} = \sum_{i=1}^{3} D_i^2 - (x - x_i)^2 + (y - y_i)^2 \tag{2.9}$$

$$\hat{x}, \hat{y} = \arg\min\left(F_{obj}\right) \tag{2.10}$$

The minimization will be carried out under the constraint such as [22]

$$\max_{i=1,2,3}(x_i - R_i) \leq x \leq \min_{i=1,2,3}(x_i + R_i) \text{ (upperbound)} \tag{2.11}$$

$$\max_{i=1,2,3}(y_i - R_i) \leq y \leq \min_{i=1,2,3}(y_i + R_i) \text{ (lowerbound)} \tag{2.12}$$

The following stages outline the aforementioned enhanced entropy-weighted approach-based RSSI localization procedure for finding position of coal mine underground miners (Figure 2.2).

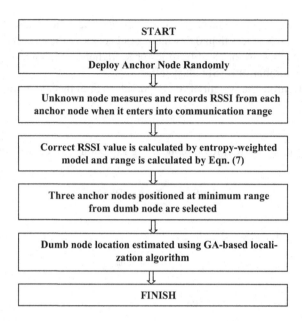

FIGURE 2.2   Flowchart for implementation of proposed algorithm.

1. After beacon nodes broadcast a positional signal, dumb nodes can connect with the beacon node. Repeat the process and keep track of the RSSI readings.
2. Formula is used to calculate the correct RSSI values for each beacon node (2.6).
3. These RSSI values are utilized in a calculation to calculate distance (2.7).
4. Distance values are chosen that are less than the communication range.
5. Finally, using a formula, identify the location of the dumb node (2.10).

2.4   RESULTS AND DISCUSSION

The experimental simulations are carried out using MATLAB® to check the validity of the proposed algorithm. We then analyze and compare its performance with traditional methods. We chose a 100 m × 100 m plane and randomly distributed 40 anchor nodes throughout the area, 20 of which are anchor nodes and 20 of which are unknown nodes. ZigBee wireless technology is utilized because it is power efficient and cheaper and has short-time delay, big network capacity, dependability, and security that suit mine needs. The path attenuation factor (n) is set to 4, and random variable X with Gaussian distribution with zero mean and four standard deviation mean is considered. As shown in Figure 2.3, the black dots represent anchor nodes, red circles represent unknown nodes true location, and blue cross represents unknown nodes estimated location. Each node has the same radius of communication, set at 30 m.

We tested the suggested methodology of entropy-weighted RSSI-based ranging model and compared it with the statistical mean, Gaussian, and Gaussian-weighted-based ranging models in terms of ranging accuracy. The entire result is done with

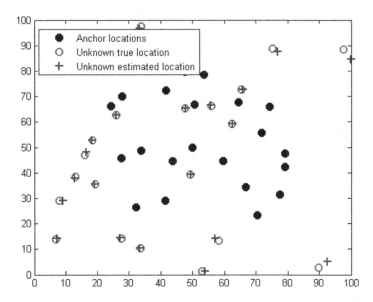

FIGURE 2.3 Location results using proposed entropy-weighted RSSI-based localization algorithm.

100 experimental simulations, and the final result is obtained by averaging all the experimental results.

Figure 2.4 depicts the graph between number of single node's RSSI and its ranging error for entropy-weighted, statistical mean, Gaussian, and Gaussian-weighted-based ranging models. The quantity of RSSI number assigned to a single node has a significant impact on range accuracy. The cost of location rises as the quantity of RSSI values rises, and positioning accuracy rises as well. It can be observed that as the quantity of RSSI values grows, the accuracy of the models improves. The ranging error diminishes quickly when the quantity of RSSI values is within 100. It can

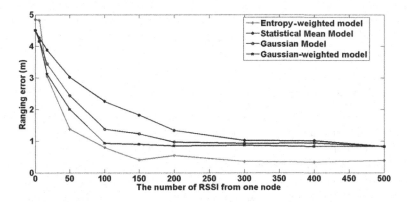

FIGURE 2.4 Relation between quantity of RSSI number of a single node and ranging error.

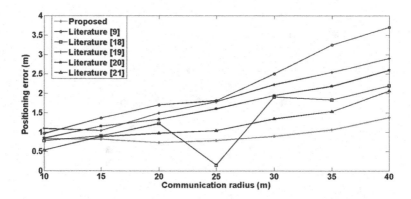

FIGURE 2.5 Comparison of proposed positioning algorithm with different algorithms based on positioning error.

be seen that the entropy-weighted model offers the best range accuracy and the most noticeable error reduction.

When there are more than 100 RSSI values, the range error decreases and becomes stable. The ranging error is relatively comparable as the RSSI number reaches 500. As per the preceding study, the number of RSSI values in this simulation should be set to 100. This saves money while maintaining great range accuracy using the entropy-weighted model.

Furthermore, after correcting the RSSI value, the positioning method in this research is compared to previously described techniques in the literature [9,18–21] in the same environment based on the positioning error [21]. Figure 2.5 depicts the simulation outcome. The position accuracy of our proposed method is the highest, as shown in Figure 2.5. Within 25 m, the algorithm of literature [18] has a reduced placement error. Its location mistake develops faster than others beyond 25 m. The algorithms other than [18] show lesser positioning accuracy, and they are, however, more stable than the literature [9]. The foregoing results reveal that the entropy-weighted model provides the highest positioning accuracy and is determined to be more stable than other methods in the case of positioning error correction. As the communication range decreases, the positioning inaccuracy reduces. The positional precision is maintained at roughly 0.8m when the communication range is 10 m.

2.5 CONCLUSION

This research examines a variety of elements that contribute to miners' positioning errors in coal mines and presents a positioning algorithm for underground coal mine workers based on RSSI. The entropy-weighted model is used to predict the correct RSSI value in coal mines because it is particularly unstable. The advantages of the entropy and weighted models are combined in this model. In a coal mine with a difficult environment, it has excellent range precision. MATLAB is used to simulate the algorithm that is proposed. According to the findings of the experiment, the entropy-weighted model was found to be the best compared to conventional algorithms. The positioning error of improved RSSI-based localization algorithm was found to be less than 0.8 m.

REFERENCES

1. Zhu, Y. and You, G.: Monitoring system for coal mine safety based on wireless sensor network. In: *2019 Cross Strait Quad-Regional Radio Science and Wireless Technology Conference (CSQRWC)*, Taiyuan, pp. 1–2 (2019).
2. Basu, S., Pramanik, S., Dey, S., Panigrahi, G. and Jana, D.K.: Fire monitoring in coal mines using wireless underground sensor network and interval type-2 fuzzy logic controller. *International Journal of Coal Science & Technology* 6(2), 274–285 (2019).
3. Fan, Q., Li, W., Hui, J., Wu, L., Yu, Z., Yan, W. and Zhou, L.: Integrated positioning for coal mining machinery in enclosed underground mine based on SINS/WSN. *Scientific World Journal*, 2014, 460415, 12 pages (2014).
4. Misra, P., Kanhere, S., Ostry, D. and Jha, S.: Safety assurance and rescue communication systems in high-stress environments: A mining case study. *IEEE Communication Magazine* 48(4), 66–73 (2010).
5. Wang, H., Roman, H.E., Yuan, L., Huang, Y. and Wang, R.: Connectivity, coverage and power consumption in large-scale wireless sensor networks. *Computer Network* 75, 212–225 (2014).
6. Liu, Z., Li, C., Wu, D., Dai, W., Geng, S. and Ding, Q.: A wireless sensor network based personnel positioning scheme in coal mines with blind areas. *Sensor* 10(11), 9891–9918 (2010).
7. Kuang, X.H., Hui-He, S. and Rui, F.: A new distributed localization scheme for wireless sensor networks. *Acta Automatica Sinica* 34(3), 344–348 (2008).
8. Singh, S. P., and Sharma, S. C.: Range free localization techniques in wireless sensor networks: A review. *Procedia Computer Science* 57, 7–16 (2015).
9. Ji, X. and Zha, H.: Sensor positioning in wireless ad-hoc sensor networks using multi-dimensional scaling. In: *Proceedings of the 23rd Annual Joint Conference of the IEEE Computer and Communications Societies (INFOCOM '04)*, Hong Kong, pp. 2652–2661 (2004).
10. Niculescu, D. and Nath, B.: Ad Hoc positioning system (APS). In: *Proceedings of the Global Telecommunications Conference (GLOBECOM '01)*, San Antonio, TX, pp. 2926–2931 (2001).
11. Liu, B.C. and Lin, K.H.: Distance difference error correction by least square for stationary signal-strength-difference-based hyperbolic location in cellular communications. *IEEE Transactions on Vehicular Technology* 57(1), 227–238 (2008).
12. Sun, G., Chen, J., Guo, W. and Liu, K. R.: Signal processing techniques in network-aided positioning. *IEEE Signal Processing Magazine* 22(4), 12–23 (2005).
13. Kumar, S. and Lobiyal, D. K.: An enhanced DV-Hop localization algorithm for wireless sensor networks. *International Journal of Wireless Networks and Broadband Technologies* 2(2), 16–35 (2012).
14. Xue, D.: Research of localization algorithm for wireless sensor network based on DV-Hop. *EURASIP Journal on Wireless Communications and Networking* 1, 1–8 (2019).
15. Anthrayose, S. and Payal, A.: Comparative analysis of approximate point in triangulation (APIT) and DV-HOP algorithms for solving localization problem in wireless sensor networks. In: *2017 IEEE 7th International Advance Computing Conference (IACC)*, Hyderabad, pp. 372–378 (2017).

16. Alrajeh, N.A., Bashir, M. and Shams, B.: Localization techniques in wireless sensor networks. *International Journal of Distributed Sensor Networks*, 9(6), 304628 (2013).
17. Zhang, J. W., Zhang, L., Ying, Y. and Gao, F.: Research on distance measurement based on RSSI of ZigBee. *Chinese Journal of Sensors and Actuators* 22, 285–288 (2009).
18. Wan, G.F., Zhong, J. and Yang, C.H.: Improved algorithm of ranging and locating based on RSSI. *Application Research of Computers* 29(11), 4156–4158 (2012).
19. Xu, J. Q., Liu, W., Zhang, Y. Y. and Wang, C. L.: RSSI-based anti-interference WSN positioning algorithgm. *Journal of Northeastern University* 31(5), 647–650 (2010).
20. Tao, W., Zhu, Y. and Jia, Z.: A distance measurement algorithm based on RSSI hybrid filter and least square estimation. *Chinese Journal of Sensors and Actuators* 25(12), 1748–1753 (2012).
21. Ge, B., Wang, K., Han, J. and Zhao, B.: Improved RSSI positioning algorithm for coal mine underground locomotive. *Journal of Electrical and Computer Engineering* 2015, 918962, 8 pages (2015). https://www.hindawi.com/journals/jece/2015/918962/
22. Peng, B. and Li, L. An improved localization algorithm based on genetic algorithm in wireless sensor networks. *Cognitive Neurodynamics* 9(2), 249–256 (2015).

3 YaraCapper – YARA Rule-Based Automated System to Detect and Alert Network Attacks

Karthik Sainadh Siddabathula,
Ramesh Kumar Panneerselvam,
Vineela Vasana, Jayavardhan Vejendla,
Mohammed Rafi, and Sree Bhanu Gummadi
VR Siddhartha Engineering College

CONTENTS

DOI: 10.1201/9781003320333-3

3.1 INTRODUCTION

Getting into the details of YaraCapper comprises some complex concepts. From the higher level of abstraction to the lower level of working, it takes a huge amount of automation work. Firstly, the scope, current YaraCapper version is compatible working with wireless networks as well as with wired networks. It's quite obvious that every enterprise network somehow runs its hops on wireless computer networks. This leveraged us to program the YaraCapper for both kinds of networks. For interfacing with a wireless network, we are using a special wireless chipset Atheros AR9271 [1].

YaraCapper is a YARA rule-based automated system to detect and alert network attacks. The preceding statement gives us a meaning that YaraCapper uses YARA rules as a validator component. In Infosec, YARA is defined as a tool to classify malware samples. Plenty of vendors like FireEye use the YARA rule to detect many attacks. At the time of writing this paper, the rolled version of YARA is 4.0.5. But the version that was integrated is 4.0.2. As we update the YaraCapper, we upfront the version of YARA too. Discussing network attacks is a never-ending topic. In our paper, we will restrict this discussion to some extent. Attacks are of two categories, active and passive. Active attacks are the kind of attacks where it involves some state change of a target, whereas passive attacks are mostly unidentifiable. The entire intention of YaraCapper is to identify any form of attack that occurs through a network. YaraCapper provides the feature of adding administrator's own rules. This pushes the YARA beyond use. Alerting the administrator is another feature of YaraCapper, the administrator mail has to be configured accordingly, such that the alerts are thrown to the admin mailbox.

Implementation of YaraCapper took place with three different sections, called engines. In our context, an engine is an abstract block consisting of complex operations that were automated. YaraCapper is comprised of three engines, named packet capture engine, YARA validation engine, and mailing alert engine. The name of the engine describes what it constitutes. Each engine plays a significant role by holding certain tasks that are meant to make the YaraCapper work. For a wireless network, the packet capture engine handles the capturing of pre-authentication, post-authentication packets, and stores the packet files, pcap, whereas for wired networks, packet capture engine captures packets from the wired interface, and there will be no pre-authentication packets in wired networks. The YARA validation engine takes the stored packet capture files, cultivates them with robust YARA rules, and flags the corresponding pcap file with the matched rule. Admin can write and place their YARA rules such that the target packet capture files can be validated with those custom rules. The mailing alert engine is what mails the flagged pcap files and their matched rule name to the administrator. YaraCapper uses a well-known mailing service, Gmail. "from," "to" mail has to be configured before the YaraCapper to be run. The "from" mail must be enabled with a third-party application password, and can be found in account settings of Gmail, and also a 2FA on. All engines are built with some components. All components are automated by YaraCapper.

Two terms might put you in chaos, pre-authentication, and post-authentication. Limiting to our context, pre-authentication packets are the one captured being unassociated with the wireless access point. Pre-authentication files may include handshakes between a SSID and a wireless station, whereas packets captured from the associated wireless access point include data packets termed as post-authentication packets.

3.2   BACKGROUND

A network interface is hardware device that could be able to send and receive packets or frames from and to the network, respectively. Sticking to 802.11i [2], a wireless network interface card supports eight different modes. Typically, these are master mode, managed mode, ad-hoc mode, repeater mode, mesh mode, Wi-Fi direct mode, TDLS mode, and monitor mode, out of which YaraCapper concerns only monitor and managed modes.

3.2.1   MONITOR MODE A.K.A RFMON MODE

Putting the wireless card into the monitor mode enables it to monitor all the traffic on a specific wireless channel. Meaning that all packets from all SSIDs are able to be captured by the wireless NIC, when put into a monitor mode. Monitor mode is different from promiscuous mode, where it captures all packets from currently associated access point. When an adapter puts into monitor mode, it gets disassociated from the SSID and starts capturing all the packets from all SSIDs. In YaraCapper, we use this mode to get pre-authentication packets, so that these can be preserved to check for wireless authentication security. Currently, YaraCapper captures pre-authentication packets, but testing wireless authentication may be implemented in future versions.

3.2.2   MANAGED MODE

Managed mode is a typical mode for many wireless NICs. In this mode, the wireless NIC can handle the packets that were sent to it. When a wireless station is connected to wireless network, the managed mode is responsible for sending and receiving packets from its associated SSID. Generally, this sort of communication happens with the help of MAC address. All traffic through that station is tunneled by managed mode only. For our YaraCapper, we use this mode to capture post-authentication packets. These captured packets are then preserved for validating with YARA rules.

3.2.3 PRE-AUTHENTICATION PACKETS

The resultant files that were captured from monitor mode are referred to be as pre-authentication packets. Pre-authentication packets may include handshake capturing or keen monitoring of specific SSID on its corresponding channel. The term is restricted to our YaraCapper context. By taking the advantage of pre-auth packets, we can also check the weakness of entry of wireless network, i.e., cracking the passphrase for wireless AP.

3.2.4 POST-AUTHENTICATION PACKETS

All the captured packets from the managed mode fall under this category. The term is also restricted to our YaraCapper context. These packets generally constitute the communication conversation between the server and a client. This communication happens with different TCP/UDP protocols such as HTTP, SSH, FTP, DNS, and many more. As of now, at the time of writing, YaraCapper generally handles TCP protocols only. YARA validation engine takes post-authentication packets as an input and flags the corresponding pcap files.

3.2.5 ATHEROS AR9271

Network market is overwhelmed with different types of wireless NICs, out of which, YaraCapper needs one. The job of choosing one is hard, as our YaraCapper is a superlative command-line tool. Although evaluating many parameters and considering many factors, we were about to choose Alfa wireless NIC – AWUS036NHA, having robust chipset known as AR9271. The Atheros AR9271 chipset supports monitor mode and packet injection as well. This is our recommendation; you can go with your choice. But keep in mind that to run out of run-time errors, you must use a chipset that supports monitor mode.

3.2.6 YARA

YARA [3] is considered as one of the most prominent tools for malware analysis. It is our honor to use it in YaraCapper as the main component. YARA provides signature-based detection to identify specific files matched with that signature. We can install YARA on Linux-based distro as well as on Windows OS. Typically, a YARA rule contains a rule name, metadata, strings, and a condition. Rule name is a title of the rule. Metadata give you details about the description, author, and any other details of the rule. Strings are actual section which involves signatures to be defined. Condition is the section that deals with Boolean expression on which the expression is evaluated for the detection to occur. YARA is very simple to write and understand as its syntax is like the language C. A demo YARA rule is shown below, and the "demoYaraCapper" rule matches when a file has hexadecimal values as given by $a.

```
rule demoYaraCapper
{
    strings:
      $a = {0A 0B 0C}
    condition:
      $a
}
```

3.3 LITERATURE SURVEY

Berger Sabbatel and Duda [4] performed an "analysis of malware network activity" (2011) that involves capturing the malware by using honeypots, analyzing the malware using sandboxed analysis tools like Norman sandbox analysis, CWSandbox analysis system, Anubis and classifying the malware using ClamAV tool. This paper helps in identifying botnet attacks easily and hence could be a great help for projects dealing with botnet attacks. Analysis of network reduces work of the network administrator. However, it only deals with the botnet detection, and usage of free tools reduces the efficiency of analyzing the malware.

Qadeer et al. [5] proposed a method for "network traffic analysis and intrusion detection using packet sniffer," (2010) by capturing traffic travelling through NIC card and filtering the traffic. This paper aids in understanding analysis and working of RTT detection and SNMP monitoring. An immediate action can be taken after the attack has been done. Packet sniffer being platform-dependent, its scope has been limited to wired networks only.

Aryeh et al. [6] performed a "graphical analysis of captured network packets for detection of suspicious network nodes," (2020) by converting the network packets into data frames and analyzing them using different python and data science techniques. Analysis of network packets will be displayed through pie charts and other visualization techniques. Using this technique, malicious packets can be identified in shortest possible time. However, malformed packets can impact the analysis of packets.

Xiaoguang and Xiaofan [7] performed a "packet capture and protocol analysis based on Winpcap" (2016), where a C++ program (written in Visual C++) is used to implement the capturing and analyzing of packets, MFC is used for encoding, and some APIs are used to check the system performance. User can easily search the data by using the highly advanced packet filter mechanism. It also gathers statistical information. However, when running across different platforms, the systems' compatibility needs to be improved and it runs only Windows OS due to Winpcap.

Prasse et al. [8] proposed a technique for "malware detection by analyzing network traffic with neural networks," (2017) that has started with developing scalable protocol to capture network flow. Now, this protocol collects network flows of known malicious traffic. Captured flow is then food for the neural network for malware detection method based on neural language and long short-term memory. Although this technique had a precise wide spectrum of dealing with HTTP malware traffic and performs network flow analysis, it was restricted to HTTP protocol. Detection occurs based on host addresses and time stamps.

Zihao and Hui [9] performed a "network data packet capture and protocol analysis on JpcapBased," (2009) that puts NIC in promiscuous mode, then captures data through the JPCAP Library. Special analysis program is used, NetMon, for protocol analysis. Promiscuous mode enabled to capture all data traffic. It is based on data and header fields and provides clear view on communication data packets through TCP/IP protocol layer. However, it uses JPCAP Library making it more platform-dependent, and is invalid to further versions of Java 2 JDK.

Li et al. [10] performed "the comparison and verification of some efficient packet capture and processing technologies," (2019) using mechanisms such as memory mapping, zero copy, and large page memory. They carried out the experiments of those high-performance packet capture technologies, and analysis occurs according to capture performance. Breakthrough points interrupted polling mechanisms.

Saavedra and Yu [11] performed a "towards large scale packet capture and network flow analysis on Hadoop," (2018) by building a tool called hcap which converts the captured packets into other format that can be understood by Hadoop, and filter and aggregate them based on the port and protocol used. Hadoop's horizontally scalable ecosystem provides better environment for processing the network captures stored in pcap files. Malformed packets can also be parsed by the hcap tool. However, it slows down the query response time of both hcap and Hadoop-pcap.

Risso and Degioanni [12] built "an architecture for high performance network analysis," (2001) where the packet generation process, even in presence of a strange behavior (maximum number of packets per second is reached with 88-bytes packets), is highly optimized. Winpcap is a tool that has high-performance packet filtering and used for many applications on Win32 platforms. The capturing process has an excellent implementation, and it outperforms the original BPF/Libpcap implementations. It improves the architecture by using circular kernel buffer instead of the original buffering architecture, and some other implementations that increase the performance. All the tests are repeated with standard (32 KB) buffers, but expected differences have not been obtained. However, a change in the size of kernel buffer does not seem to have influenced the performance of the capture process.

Neupane et al. [13] conducted a "next generation firewall for network security: A survey," (2018) based on the vulnerabilities of traditional firewalls (like AET, web application attacks, data-focused attacks, etc.). NextGen firewall overcomes all these vulnerabilities by deep packet inspection. Different vendors will have different features of NextGen firewall. However, this paper only focused on technology implementation and advanced features of NGFW, and failed in mentioning the pros and cons over various kinds of firewalls.

Zhang and Zulkernine [14] proposed an "anomaly based network intrusion detection with unsupervised outlier detection," (2006) technique that reduces false positive rate with a desirable detection rate. This approach is using the KDD'99 dataset which is the mostly used dataset to evaluate anomaly detection. The problem with supervised anomaly-based IDS is small change in the network services or environment leads to high false positive rate, and this approach is not suitable for real-world network environment. Unsupervised approach over-comes all these limitations by using the outlier detection algorithm of random forests providing higher detection rate and less false positive rate. Problem with unsupervised systems is the performance has been observed to be reduced with increasing attack connections, i.e., DoS attacks can sabotage the unsupervised anomaly detection system.

Dabir and Matrawy [15] performed "bottleneck analysis of traffic monitoring using Wireshark," (2007). Wireshark is a tool that analyzes network traffic in the real-time environment, and it is one of its executables Dumpcap writes packets to the hard disk; concurrently, the GUI component read the stored packets from the disk to process. But when dealing with high packet rates, a single-threaded application will not work. So, they proposed multi-threaded solution that allows the capturing application to support buffering independently of Libpcap, thus increasing response time of the application.

Witzke [16] conducted a survey on "computer network security: then and now," (2016). This paper presented a review on the state of computer and network security in 1986, about today's security environment and what improvements are made. It pro-vides a comprehensible analysis about different types of threats and their evolution since 1986. However, it only focused on the threats and attacks but did not mention anything about the history of the preventive measures.

Nespoli et al. [17] proposed "cyber protection in IoT environments: A dynamic rule-based solution to defend smart devices" by improving the security character-istics through adding different modules (DRL, VDS, EPM, and MM) to COSMOS approach, which is a shield for IoT devices in a network. Instead of using the single-module COSMOS approach, they modularize the components so that they will be extensible and simple to maintain in the future. However, the processing power of IoT devices is low, which reduces the performance.

Brengel and Rossow [18] proposed "YARIX: scalable YARA-based malware intelligence" uses YARA rules to match the pattern of the malware present in the files, and it uses the n-gram index technique to easily find out the malware files in large organizations. It reduces disk foot printing and increases performance. However, it is difficult to find out the n-value in the n-gram index technique, because different n-values will result in a change in the performance.

3.4 PROPOSED SYSTEM

It is been a storm for us to craft a new architecture that gives a life to the YaraCapper. We need our architecture to be loosely coupled such that it can be used for the inte-gration of other tools as well. The loosely coupled nature of our architecture also helps in easy debugging by eliminating all tedious things. Initially, when design is still in mind, we are about to program the YaraCapper with a single engine and with single source file. But we need our architecture to be easy plugged so that anyone, in future, who contributes their time to project, should understand its internal working. This kind of architecture development not only helps in debugging but also makes upgrade very smoother (Figure 3.1).

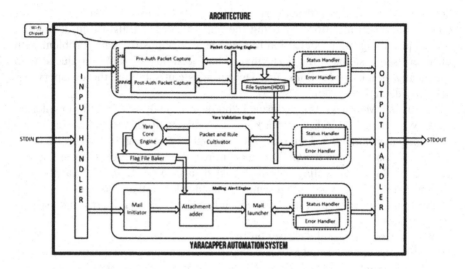

FIGURE 3.1   Architecture of YaraCapper.

3.4.1 WI-FI CHIPSET

Already been said, any Wi-Fi card that supports monitor mode is sufficient to handle the working of YaraCapper. We recommend to use the Alfa AWUS036NHA card. The most important factor is that the packet capture engine and mailing alert engine utilize the Wi-Fi adapter to capture packets and for sending mail to admin, respectively. Hence, a stable Internet connection may require alerting the administrator. We use the Wi-Fi chipset to perform wireless validations.

3.4.2 STDIN – STANDARD INPUT

Yes, YaraCapper is an automated system, but there should be some parameters to regulate the working of any automated system. Reading wireless card point, BSSID, channel, number of post-auth packets to be captured, packet files path, rule files path, to mail address are done through STDIN. Some inputs are handled by YaraCapper itself, but some are read from the user.

3.4.3 STDOUT – STANDARD OUTPUT

As YaraCapper is a command-line utility, the status and error messages are written to STDOUT. We didn't use STDERR as of now. Statements return to the STDOUT include "success status," "generic status," "error status," represented by "[+]," "[*]," and "[−]" with color "green," "yellow," and "red," respectively. This kind of representation helps the administrator to read the output wisely.

Three engines are the core reason to drive the YaraCapper. Each engine has its dependencies, and it will start when they meet dependencies. The status handler and error handler components work the same for all engines as mentioned in Sections 4.2 and 4.3.

3.4.4 PACKET CAPTURE ENGINE

First engine of YaraCapper is responsible for capturing packets in both wired and wireless networks. There were some dependencies on which the engine runs. Aircrack-ng and tshark are the tool dependencies of this engine. The first and foremost that a packet capture engine does is to check for the log file, if no log file exists, it creates one. The next thing is to search whether Aircrack-ng and tshark are installed or not. Aircrack-ng is for wireless network capture, and tshark is for wired and wireless as well. If the dependencies are unsatisfied, YaraCapper exits. Remember this dependency check is only for packet capture engine. Other engines may have other dependencies. Packet capture engine contains some versatile components, namely pre-auth capture, post-auth capture, status handler, and error handler.

There is a preliminary check for Wi-Fi card. It is obvious for a system to have more than one Wi-Fi card; this engine also prompts whether to use another Wi-Fi card instead another other than detected one. We read wireless card device id from user input if in case the detected one is not the desired one for the admin. The same applies to wired interfaces as well. If another wired interface is preferred than the default one, admin may give it at the time of prompt.

3.4.4.1 Pre-Authentication Capture

This component takes the help of the airmon-ng tool to keep the wireless card into monitor mode. Next, YaraCapper utilizes airodump-ng to dump the surrounded Wi-Fi networks, out of which we have to choose one by giving respective BSSID and channel. YaraCapper now turns to capture the pre-auth packets of given BSSID, CH. These captured packets are driven to the filesystem. A handshake might be captured during this time, as it depends on a Wi-Fi station tends to connect. At last, this component also puts back the Wi-Fi card into the managed mode such that it can be used for the post-auth capture component. Restarting the network service is the final task of this component. All output is sent to the output handler as well as errors if any. The pre-auth capture component is only for wireless mode.

3.4.4.2 Post-Authentication Capture

As the name describes, it captures post-authentication packets. This component directly accesses managed mode interface of the Wi-Fi card in wireless mode and the Ethernet interface in wired mode for capturing the packets. tshark is the core tool that we used to program this component. tshark dependency is checked to proceed further. It is now a tedious task to determine how many packets are to be captured. YaraCapper leaves this decision to the admin by prompting for the input of the number of packets. Post-authentication packets capturing starts, and the main point to notice that the filename of that pcap is in the format of YYMMDDHHMMSS. pcapng. As this format avoids some sort of namespace conflict, error occurrences are rendered to the output handler.

3.4.5   YARA Validation Engine

Considering any automation system, the involvement of dependencies results in playing a crucial role, but those are needed. The status handler and error handler deal with the outputting status of YaraCapper and show errors, respectively. The others are the components that include packet and rule cultivator, YARA core engine, and flag file baker.

3.4.5.1   Packet and Rule Cultivator

In our context, the term Cultivator refers to handling the flow. After the successful capturing of post-authentication packets, those have to be validated against the YARA rules. The packet files and rule files become the input for this component. And YaraCapper coins it as a flow. Strictly, Packet and Rule Cultivator component deals with the flow of captured post-auth packet files and YARA rule files. Resultants of the cultivator are directed to the YARA core engine component. In other words, the YARA core engine is a sub-component of the Cultivator component. The beauty of this cultivator is that it takes all pcap files and all rule files at once and does broadcast flagging with the YARA core engine. This means that the cultivator picks one pcap file and sends it to the YARA core engine to validate it with all YARA rules, each rule at once. "one-to-all" term may fit this kind of process. This continues until all pcap files with all rules are directed to the YARA core engine. Status can be seen and any errors are redirected to the error handler.

3.4.5.2   YARA Core Engine

The actual YARA compiler is referred to as the YARA core engine, in YaraCapper. The current version of the YARA core engine is 4.0.2, at the time of development. Another alternative is installing YARA API. The main function of this YARA core engine is to check the pcap file with the rule given. These inputs are fetched from the preceding components called Packet and Rule Cultivator. If the pcap file has satisfied with any rule, then the pcap file is flagged and pushed to the flag file baker component.

3.4.5.3   Flag File Baker

The flagged pcap files are needed with a little bit of roasting. Roasting information includes what to deliver in the mail. In other words, Baker marks pcap files to which the rules are satisfied. Typically, the marking is done with the corresponding rule name. And the respective pcap files are then sent to the third engine, mailing alert engine. If there are no pcap files that got flagged, then the YaraCapper exits successfully without sending mail, and unnecessary alerts may fill the admin's inbox. All status and errors are handled.

3.4.6   Mailing Alert Engine

The rearmost engine is known to be the lowest job engine as the whole responsibility is to send the mail of given information. The obvious metrics to be needed for this engine to work are "from mail," password, "to mail." Alert engine is invoked only when there are flagged pcap files. Keeping the dependencies, aside, a basic requirement of this engine is a stable Internet. The reason for this is it gets connected to the SMTP server to send mail. The dependencies include the SMTP library. The SMTP server we are using is Gmail, which becomes our mail server. As always without dependencies, the engine won't work. Errors are handled along with statuses. Components of this engine are built robust due to security reasons. Mail initiator, attachment adder, and mail launcher are the components of mailing alert engine.

3.4.6.1 Mail Initiator

Mail initiator typically requires the "from mail," and its password, "to mail." This is taken from the input handler. As already discussed in Section 4.2, these inputs are handled by YaraCapper itself, but they have to be mentioned statically traversing the script at least once. And the initiated entity is given to the next immediate component for further action to take place. As the YaraCapper version increases, we also improve the security to be given for mail and password. The current security scheme follows a third-party application password feature of Gmail. Adding subject, the main body will also get configured by the mail initiator component. The outcome of this component can be known as the mail initiator entity.

3.4.6.2 Attachment Adder

The results of the flag file baker component and outcome of the mail initiator component are fabricated together to send the mail. This includes flagged pcap files and details such as "from mail," subject, content, "to mail." Attachments are added with default MIME type as "application/vnd.tcpdump.pcap". The main type is generally the "application" whereas the subtype is "vnd.tcpdump.pcap". A neatly crafted mail object will be the outcome of the attachment adder component, then forwarded to the mail launch component.

3.4.6.3 Mail Launch

The foremost job of the mail launch component is to establish a connection to the SMTP server at smtp.gmail.com on 465 port. Port 465 enables us to send mail securely, and further performs login into the "from mail" account with the configured password. If there is any error while logging into the account, then this may drag an error and troubleshooting should happen manually. Most times the password becomes the complex area that error may raise. If everything goes well, then the mail is sent to the desired "to mail" administrator.

Figure 3.2 represents the flow of how YaraCapper automates the entire process of capturing the packets, validating them with YARA, and mailing alerts.

The current, well-crafted architecture is subjected to version 3.0 of YaraCapper. As the version gets upgraded, additional components may be added.

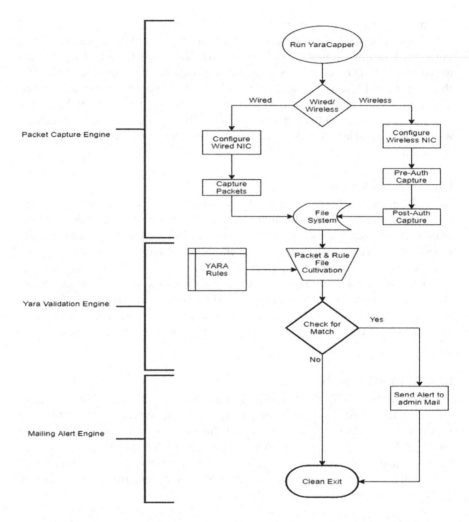

FIGURE 3.2   Flowchart of YaraCapper.

3.5   IMPLEMENTATION

In this wide section, we are going to look at the platform we chose to develop the YaraCapper, working of the YaraCapper, the testing phase of YaraCapper, the ways that we deliver YaraCapper to the enterprise, and contribution to YaraCapper.

3.5.1   PLATFORM AND LANGUAGE USED

Typically, we refer to the platform as the operating system that the development is to opt for. After some level of groundwork, the entire YaraCapper turned to build on a Linux track. Specifically, we picked up Kali-Linux to start giving life to YaraCapper, and periodically we also opt for Ubuntu distribution to work on. The former engine, packet capture engine, was entirely developed in Kali-Linux. Kali-Linux provides great penetration testing tools such as Aircrack-ng, and this makes it easier to organize the dependency calls within the engine itself. The latter two engines, YARA validation engine and mailing alert engine, were developed in Ubuntu. Although latter engines can also be developed in Kali-Linux, to make the YaraCapper work on many flavors of Linux, we opted for Ubuntu as well. We also chose Debian distribution, but that is for the other section.

Talking about the language, we have used bash scripting and python. At the time of development of YaraCapper, the version of bash is 5.1.4, whereas the python version is 3.9.1. The straightforward reason for picking these two languages is they are apt for automation works. This makes YaraCapper be a command-line utility.

3.5.2 TESTING OF YARACAPPER

Core level API to high level abstraction is very far tested in terms of working. But, it is always known that a bug lies behind what we see. Feel free to communicate with us to report bugs, if any. We have performed many regression tests to YaraCapper on Ubuntu, and Kali-Linux. Testing is also done on YaraCapper deliverable medium – Debian.

3.5.3 DELIVERING THE YARACAPPER

The YaraCapper was started to serve enterprise networks, SOHO networks, and many. During the development, we decided to give two different ways to drag YaraCapper into the network:

3.5.3.1 Manual Install

One can directly visit the GitHub page of YaraCapper and follow the instructions to install YaraCapper. This is tedious to deal with dependencies but more flexible when installed in correct manner.

3.5.3.2 Debian Ready VM

We decided to install YaraCapper in Debian OS and to give a readymade virtual machine such that deployment in a virtualized environment is done easily. This also eliminates the dependencies installation. The cause to choose Debian is that every enterprise uses Debian as most common Linux flavor, apart from CentOS and RedHat.

3.5.4 CONTRIBUTION TO YARACAPPER

YaraCapper welcomes security community people to contribute new features. Yes, this means our YaraCapper is an open-source software and released under GNU General Public License v3. Sticking to the license anyone utilizes the YaraCapper.

3.6   RESULTS AND ANALYSIS

3.6.1   RESULTS

Before running the YaraCapper, we need to set the from and to email along with a third-party application password. Configure third-party application password access on admin's Gmail account. Now carefully edit the "YaraValidateMail.py" file to add "from mail," "to mail," and password. Remember, the admin needs to give a third-party password that has been configured (Figure 3.3).

Step 1: Install all the dependencies by running "install-dependencies" script with root privileges.

Step 2: Now YaraCapper is ready to run. To display the menu, issue "./YaraCapper. sh" (Figure 3.4).

Step 3: Wireless network mode: To run YaraCapper for wireless network. Issue: "./YaraCapper.sh -w". YaraCapper starts with printing the banner and sets the script to serve wireless network (Figure 3.5).

```
# Assigning sender and reciever mail addresses
frommail = "xyz@gmail.com" # change from mail here
passwd = "" # add passwd here
tomail = "" # add admin mail here
```

FIGURE 3.3   Edit credentials.

```
root@kali:/opt/YaraCapper# ./YaraCapper.sh
Usage: ./YaraCapper.sh <option>
Options:
-h    : help
-e    : Ethernet Capture
-w    : Wireless Capture
-b    : Prints Banner
```

FIGURE 3.4   YaraCapper help menu.

FIGURE 3.5   Banner of YaraCapper.

Step 3.1: Next, YaraCapper checks for dependencies that it relies on. If any dependencies will not satisfy, run the script described in Step 1. After dependency check, it searches for the default wireless NIC card. A default will be considered initially. The administrator can change the wireless card of their choice in the next step (Figure 3.6).

Step 3.2 (optional): At this point, it prompts for a change of wireless card. If admin wishes to change, then they need to give the new wireless interface point. It validates the given interface. If found, it continues. Else, YaraCapper exits as shown in Figure 3.7.

Step 3.3: The wireless adapter will be configured accordingly to reduce runtime exceptions. Monitor mode will also be enabled as shown in Figure 3.8. It asks us to press "q" two times when we see Wi-Fi networks.

FIGURE 3.6 Dependency checking and default interfacing.

FIGURE 3.7 Interface validation.

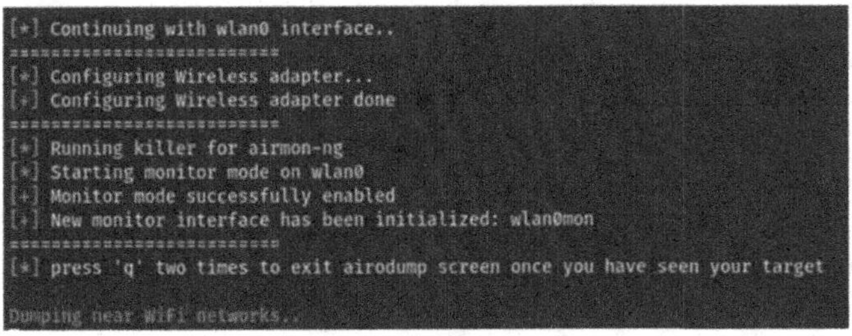

FIGURE 3.8 Configuration of adapter and status of monitor mode.

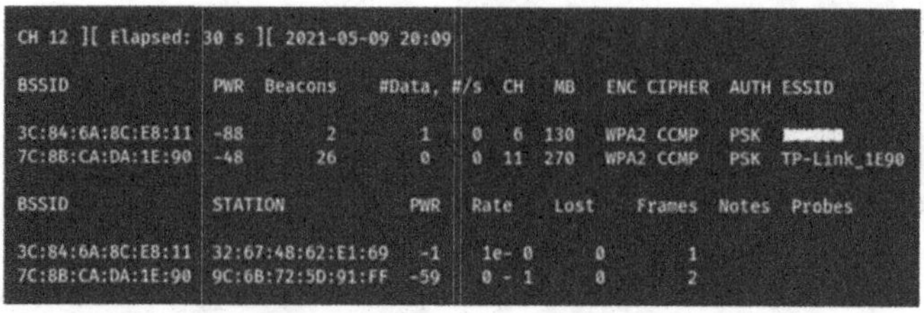

```
CH 12 ][ Elapsed: 30 s ][ 2021-05-09 20:09

BSSID              PWR  Beacons    #Data, #/s  CH   MB   ENC  CIPHER AUTH ESSID

3C:84:6A:8C:E8:11  -88        2        1    0   6  130   WPA2 CCMP   PSK  ███████
7C:8B:CA:DA:1E:90  -48       26        0    0  11  270   WPA2 CCMP   PSK  TP-Link_1E90

BSSID              STATION           PWR   Rate    Lost    Frames Notes Probes

3C:84:6A:8C:E8:11  32:67:48:62:E1:69  -1   1e- 0     0        1
7C:8B:CA:DA:1E:90  9C:6B:72:5D:91:FF  -59  0 - 1     0        2
```

FIGURE 3.9 Nearby Wi-Fi networks.

```
Enter the target BSSID: 7C:8B:CA:DA:1E:90
Enter the channel of BSSID: 11
```

FIGURE 3.10 BSSID and channel input.

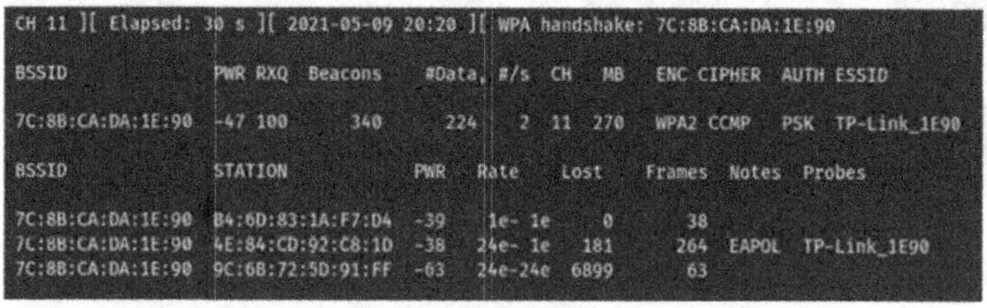

```
CH 11 ][ Elapsed: 30 s ][ 2021-05-09 20:20 ][ WPA handshake: 7C:8B:CA:DA:1E:90

BSSID              PWR RXQ Beacons    #Data, #/s  CH   MB   ENC  CIPHER AUTH ESSID

7C:8B:CA:DA:1E:90  -47 100     340      224   2  11  270   WPA2 CCMP   PSK  TP-Link_1E90

BSSID              STATION           PWR   Rate    Lost    Frames Notes Probes

7C:8B:CA:DA:1E:90  B4:6D:83:1A:F7:D4  -39  1e- 1e     0        38
7C:8B:CA:DA:1E:90  4E:84:CD:92:C8:1D  -38  24e- 1e   181      264  EAPOL  TP-Link_1E90
7C:8B:CA:DA:1E:90  9C:6B:72:5D:91:FF  -63  24e-24e  6899       63
```

FIGURE 3.11 Pre-authentication packet capture – handshakes.

Step 3.4: YaraCapper now shows all nearby Wi-Fi networks, out of which we need to select one. Once we see our target network, hit "q" two times (Figure 3.9).

Step 3.5: It asks for the BSSID and channel. Enter them as you see (Figure 3.10).

Step 3.6: Now the pre-authentication packets will be captured from the given BSSID and channel. pre-authentication packets may include handshakes. After giving it sometime, hit "q" two times to exit pre-auth capture (Figure 3.11).

Step 3.7: All pre-auth packets are to be written to the pre-auth folder for future needs, if any. Managed mode will be enabled on wireless NIC and post-auth capture starts. And it asks for the number of packets to be captured. Enter the number. Default was set to 10 (Figure 3.12).

Step 3.8: After the post-auth packets are captured, it shows a success note (Figure 3.13).

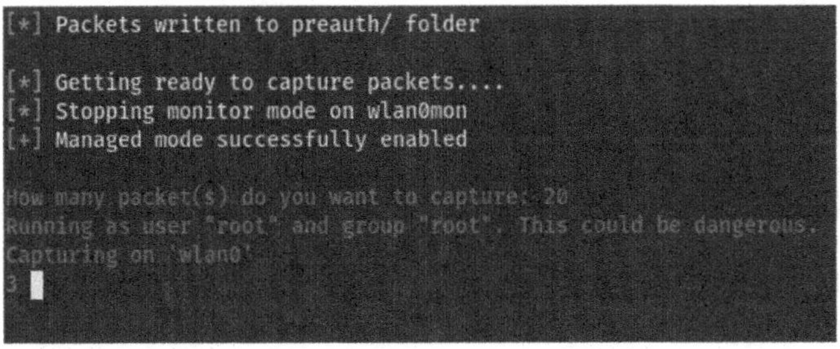

FIGURE 3.12   Status of managed mode and post-auth initiative.

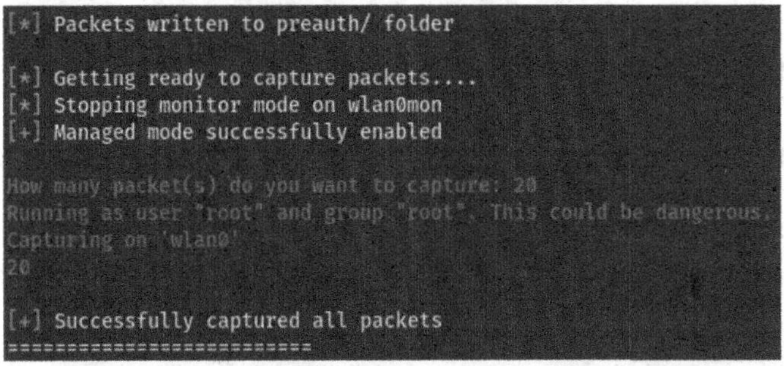

FIGURE 3.13   Post-authentication packet capture – success note.

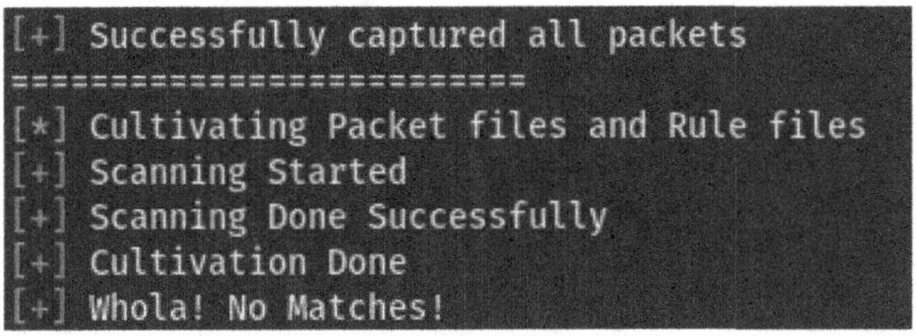

FIGURE 3.14   YARA validation and mail status.

Step 3.9: Now these post-auth packets are fed to YARA validation engine along with YARA rules for validation. If YaraCapper has a match, it sends mail. Here, there are no matches (Figure 3.14).

Step 4: Wired network: To run YaraCapper for wired network. Issue: "./YaraCapper.sh -e". YaraCapper starts with printing the banner and sets the script to serve wired network.

Step 4.1: YaraCapper now checks for dependencies and searches for default interface. If dependencies will not be satisfied, then run the script described in Step 1. Admin can change the default interface to their choice (Figure 3.15).

FIGURE 3.15 Dependency checking and default interfacing.

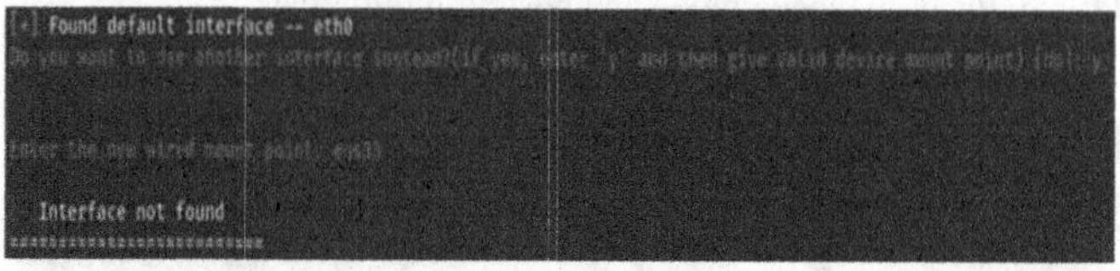

FIGURE 3.16 Interface validation.

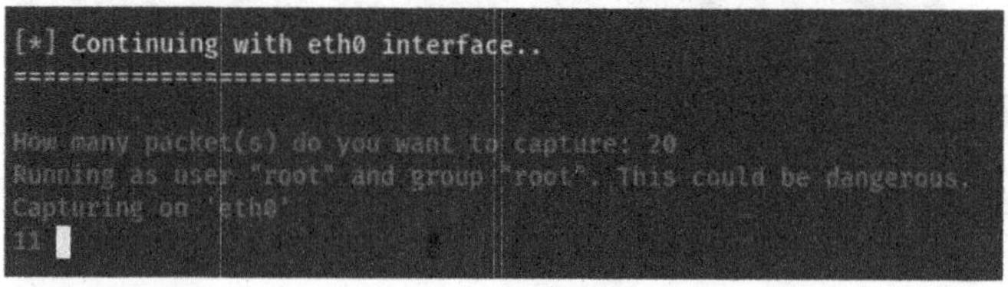

FIGURE 3.17 Wired packet capture.

Step 4.2 (optional): A new interface, if admin wishes to change, should be given. It validates the interface. If found, it continues. Else, YaraCapper exits (Figure 3.16).

Step 4.3: YaraCapper asks for number of packets to be captured. Default was set to 10. Enter the number of packets (Figure 3.17).

Step 4.4: It shows the success of packet capture. And cultivation of packet files with rules will be started. YARA validation engine validates and checks for matches. If there is any hit, YaraCapper sends mail; otherwise, it doesn't send any mail.

3.6.2 ANALYSIS

Let us see a conditional representation of how YARA validation engine handles packet capture files and rules.

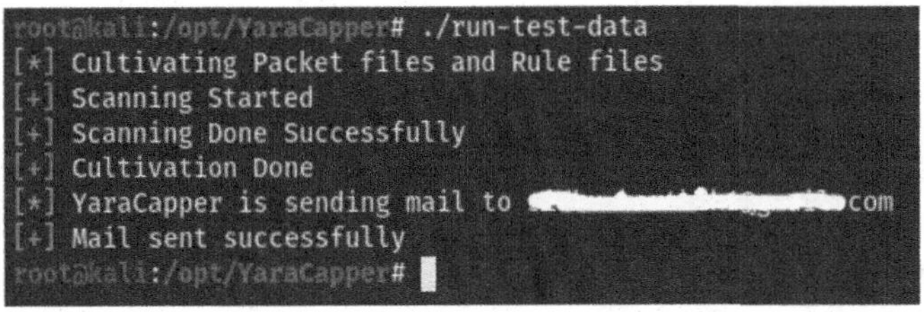

FIGURE 3.18   YARA validation and mail status for test data.

P1, P2, P3, ... Pn is the series of packet capture files. Y1, Y2, Y3, ... Yn is the series of YARA rules. Then,

$$P1(Y1) \rightarrow P1.Y1$$

$$P1(Y2) \rightarrow P1.Y2$$

$$P1(Y3) \rightarrow P1.Y3$$

$$... \rightarrow ...$$

$$P1(Yn) \rightarrow P1.Yn$$

If P1 gets satisfied with respect to rule Y1, then the resultant will be P1.Y1......... P1 packet file concatenated with Y1 rule and so on same for P2, P3, ... Pn. In other words, we can also define as follows:

pcap-file((rule-file)) –on-satisfy—> send-mail(pcap-file.rulename)

YaraCapper is as strong as the rules it has. It doesn't matter to YaraCapper about the size of the packet capture file. In the results subsection, we have given only 20 packets to be captured and in which there might be no malicious content available. For this reason, we generated some malicious traffics impersonating an attacker's perspective and captured the traffic.

Now let us feed these malicious packet capture files to the YARA validation engine directly and check for results.

In Figure 3.18, the test data are sent to YARA validation engine, and it yielded a match. This results in sending an alert mail to the administrator mail configured.

The test data contain malicious traffics related to CyberGate Malware, Satan Mutexes Malware, Solarbot, and crimepack_jar. The following figure shows the mail that has been sent by the YaraCapper to the admin mail. The mail has text describing the infection about the pcap files and its matched rules, and also it contains attachments for sending them to forensic lab as well (Figure 3.19).

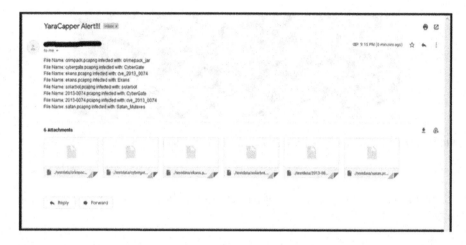

FIGURE 3.19   Mail inbox.

Validating the packets with YARA rules and its time factor analysis is detailed in Table 3.1. And the test data validation and its time consumption analysis are also portrayed in Table 3.2.

The above given validation times in Tables 3.1 and 3.2 may differ. The obvious reason is that validation is highly dependent on the processing power of the machine on which the YaraCapper is running on. But, it will be the closer of what we

TABLE 3.1
YaraCapper Time Factor Analysis

pcap File	Number of Packets Validated Against 22 Rules	pcap File Size (kB)	Validation Time (seconds)
File-50.pcapng	50	16	1.04
File-100.pcapng	100	24	1.11
Filc-200.pcapng	200	30	1.31
File-400.pcapng	400	95	1.28
File-800.pcapng	800	143	1.21
File-1000.pcapng	1,000	410	1.35

TABLE 3.2
YaraCapper Test Data Analysis

Test pcap File	Packets Validated Against 22 Rules	Matched Rule	Validation Time (seconds)
Testdata1.pcapng	100	Cyber Gate	0.5
Testdata2.pcapng	150	Satan Mutexes	0.78
Testdata3.pcapng	200	Solar bot	0.8
Testdata4.pcapng	200	Crimepack_jar	1.0

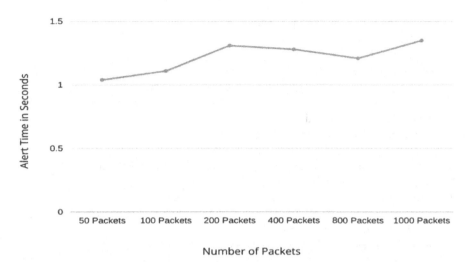

FIGURE 3.20   Line graph representation of Table 6.2.1.

have mentioned. The line graph representation of Table 3.1 is given in Figure 3.20 where the alert time is between 1 and 1.5 seconds for 50 to 1000 packets. The analysis took place with the aid of a typical Linux computer (Figure 3.20).

3.7   DISCUSSION

Currently, YaraCapper has its limitations such as not having the more efficient way of delivery, handling run-time input data. Not limited to these. Generating rules dynamically based on behavior of network is also the feature that could be considered as a future work. The delivery of YaraCapper may be made efficient when we can make use of Docker. Providing extra security to the admin mail disclosure also fits good metric. Evaluating the attack rate of the network can be an add-on. Any other features are welcome.

3.8   CONCLUSION

It is difficult to craft words to describe the YaraCapper, we strived to cut the chase. To conclude, YaraCapper captures packets and checks for malicious content using YARA rules and simply delivers the pcap files to the administrator's mail. Apart from installing the YaraCapper, we also offer a readymade virtual machine made with Debian OS containing YaraCapper. Contribution to the YaraCapper can be done with subjected to GNU General Public License v3.

Author Contributions:
    Conceptualization – Karthik;
    Methodology – Karthik, Vineela;
    Software – Karthik, Vineela, and Jayavardhan;
    Validation – Ramesh Kumar, Vineela, Rafi;

Formal analysis – Jayavardhan, Sailaja, Rafi;
Investigation – Rafi, SreeBhanu;
Resources – Karthik, Vineela;
Data curation – Jayavardhan, SreeBhanu;
Writing – original draft preparation – Karthik, Vineela;
Writing – review and editing – Karthik, Vineela, and Jayavardhan; Visualization – Vineela, Karthik, Rafi, and SreeBhanu;
Supervision – RameshKumar, Rafi;
Project administration – Sailaja, Rafi, Karthik.
All authors have read and agreed to the published version of the manuscript.

Funding: This research received no external funding.

Acknowledgments: We sincerely thank our Department of Computer Science and Engineering of Velagapudi Ramakrishna Siddhartha Engineering College for giving this cheering project to initiate.

Conflicts of Interest: The authors declare no conflict of interest.

REFERENCES

1. M. Vanhoef and F. Piessens, "Advanced Wi-Fi attacks using commodity hardware," *Proceedings of the 30th Annual Computer Security Applications Conference (ACSAC '14)*, Association for Computing Machinery, New York, NY, USA, 2014, pp. 256–265. doi: 10.1145/2664243.2664260.
2. B. P. Crow, I. Widjaja, J. G. Kim and P. T. Sakai, "IEEE 802.11 wireless local area networks," *IEEE Communications Magazine*, vol. 35, no. 9, pp. 116–126, 1997, doi: 10.1109/35.620533.
3. N. Naik, P. Jenkins, N. Savage, L. Yang, K. Naik and J. Song, "Embedding fuzzy rules with YARA rules for performance optimisation of malware analysis," *2020 IEEE International Conference on Fuzzy Systems (FUZZ-IEEE)*, 2020, pp. 1–7, doi: 10.1109/FUZZ48607.2020.9177856.
4. G. Berger-Sabbatel and A. Duda (2011). Analysis of malware network activity. In: Dziech A., Czyżewski A. (eds) *Multimedia Communications, Services and Security. MCSS 2011. Communications in Computer and Information Science*, vol 149, pp. 207–215. Springer, Berlin, Heidelberg.
5. M. A. Qadeer, A. Iqbal, M. Zahid and M. R. Siddiqui, "Network traffic analysis and intrusion detection using packet sniffer," *2010 Second International Conference on Communication Software and Networks*, Singapore, 2010.
6. F. L. Aryeh, B. K. Alese and O. Olasehinde, "Graphical analysis of captured network packets for detection of suspicious network nodes," *2020 International Conference on Cyber Situational Awareness, Data Analytics and Assessment (CyberSA)*, Dublin, Ireland, 2020.
7. A. Xiaoguang and L. Xiaofan, "Packet capture and protocol analysis based on Winpcap," *2016 International Conference on Robots & Intelligent System (ICRIS)*, Zhangjiajie, 2016.
8. P. Prasse, L. Machlica, T. Pevný, J. Havelka and T. Scheffer, "Malware detection by analysing network traffic with neural networks," *2017 IEEE Security and Privacy Workshops (SPW)*, San Jose, CA, 2017.
9. S. Zihao and W. Hui, "Network data packet capture and protocol analysis on JPCAP-based," *2009 International Conference on Information Management, Innovation Management and Industrial Engineering*, Xi'an, 2009.

10. J. Li, C. Wu, J. Ye, J. Ding, Q. Fu and J. Huang, "The comparison and verification of some efficient packet capture and processing technologies," *2019 IEEE International Conference on Dependable, Autonomic and Secure Computing, International Conference on Pervasive Intelligence and Computing, International Conference on Cloud and Big Data Computing, International Conference on Cyber Science and Technology Congress*, Fukuoka, Japan, 2019.

11. M. Z. N. L. Saavedra and W. E. S. Yu, "Towards large scale packet capture and network flow analysis on Hadoop," *2018 Sixth International Symposium on Computing and Networking Workshops (CANDARW)*, Takayama, 2018.

12. F. Risso and L. Degioanni, "An architecture for high performance network analysis," *Proceedings of Sixth IEEE Symposium on Computers and Communications*, 2001, pp. 686–693, doi: 10.1109/ISCC.2001.935450.

13. K. Neupane, R. Haddad and L. Chen, "Next generation firewall for network security: A survey," *SoutheastCon 2018*, 2018, pp. 1–6, doi: 10.1109/SECON.2018.8478973.

14. J. Zhang and M. Zulkernine, "Anomaly based network intrusion detection with unsupervised outlier detection," *2006 IEEE International Conference on Communications*, 2006, pp. 2388–2393, doi: 10.1109/ICC.2006.255127.

15. A. Dabir and A. Matrawy, "Bottleneck analysis of traffic monitoring using Wireshark," *2007 Innovations in Information Technologies (IIT)*, 2007, pp. 158–162, doi: 10.1109/IIT.2007.4430446.

16. E. L. Witzke, "Computer network security: Then and now," *2016 IEEE International Carnahan Conference on Security Technology (ICCST)*, 2016, pp. 1–7, doi: 10.1109/CCST.2016.7815710.

17. P. Nespoli, D. Díaz-López and F.G. Mármol, "Cyberprotection in IoT environments: A dynamic rule-based solution to defend smart devices," *Journal of Information Security and Applications*, vol. 60, 2021, p. 102878.

18. M. Brengel and C. Rossow, "YARIX: Scalable YARA-based malware intelligence," *USENIX Security Symposium*, August 2021.

4 Describing Nature-Inspired Networking and Real-Time Task Execution
A Theoretical Study

Rashmi Sharma
Shri Vaishnav Institute of Computer Application
Shri Vaishnav Vidyapeeth Vishwavidyalaya

Kaustubh Lohani
Stevens Institute of Technology

Prajwal Bhardwaj
University of Texas

CONTENTS

4.1   INTRODUCTION

For the completion of numerous tasks or message transmission, connectivity in-between processors are important. For a well-timed, sequential event execution, different connectivity (paths) between source and destination requires various algorithms. Similarly, on-time execution of real-time tasks [1,2] in distributed systems [3,4] helps in maintaining the overall system performance. This article will explain the role of processor's connectivity concerning task execution performance in distributed systems. To explain the perception, this work has been divided into three modules: (a) connectivity topologies of processors in distributed systems, (b) real-time scheduling algorithms, and (c) topological consideration in network distribution, based on deadline achievement.

DOI: 10.1201/9781003320333-4

The first module is the centric point of this paper where connectivity patterns of processors of distributed systems will explain. Here, our motive is to check the finest connectivity in-between processors with the help of nature-inspired networks [5]. Nature-inspired networks are those various networks that we encounter in nature, society, or day-to-day life. This nature-inspired network belongs to network science an emerging discipline of science [6]. Following three networks, we have approached for the fulfillment of this awareness: (a) regular network, (b) random network, and (c) scale-free network. Real-Time scheduling algorithms will use to schedule tasks on participant processors of the respective network.

The second module is a real-time scheduling algorithm [7] that uses to schedule real-time tasks based on their deadline. Real-Time system (RTS) [8] define a system that works on temporal behavior like arrival time of a task, worst-case execution time, its period, and deadline. To execute these tasks, first scheduling is done with the help of some scheduling algorithms. These scheduling algorithms are of two types [7,9]: static and dynamic. In static, once priority assigned to tasks will same throughout, whereas in dynamic, tasks priority can be changed after the arrival of another task. Rate monotonic scheduling algorithm is one of the examples of static priority algorithm, and earliest deadline first (EDF) is dynamic priority algorithm example. In this document, these two scheduling algorithms will be used for task scheduling purposes.

In the final module, the above-stated components will integrate where multiple real-time tasks will arrive in the system, and said scheduling algorithms will schedule these tasks on contributed processors of a distributed system. In case of overloading, tasks will migrate to their connected processor (based on connectivity and processor utilization). Now, here connectivity between processors helps in the timely execution of migrated tasks. Hence, linking between processors is responsible for the deadline achievement of tasks in distributed systems. This will help to decide the optimum connectivity among processors for the successful execution of real-time tasks.

Further, this paper explains the above modules in the background section. The third section will explain the proposed concept with a mathematical model of stated nature-inspired networks. Additionally, the result and discussion sections elaborate benefits of using the finest distributed systems for tasks execution. Lastly, the conclusion and future scope encapsulate the article.

4.2   BACKGROUND AND PRELIMINARIES

This section explains the types of processors' connectivity with real-time scheduling algorithms for task scheduling purposes. A distributed system is a connection of processors with each other, and with the help of graph [10,11], the generalized representation of connectivity can show this connectivity. So, before explaining the features of different three networks, let us discuss some basic graph terminologies in Table 4.1

Following are three types of network strategy, we shall use for the accomplishment of the proposed work:

a. Regular network
b. Random network
c. Scale-free network

TABLE 4.1
Basic Terminologies of Graph

S. No.	Terminologies	Definition
1	Graph	The graph is a picture that represents the relationship between two vertices/nodes with the help of an edge/link between them. $G(N,L)$ is a mathematical representation of a network where N is a node and L is a relationship of respective node N with other nodes.
2	Nodes	Nodes are the vertices of a network/graph, that is, N.
3	Links	Links are the connection in-between nodes or edges between vertices of a graph. L is the representation of links.
4	Degree	Degree defines the number of connections respective nodes having. Degree (i) represents the total connections of node i, where $i = 1,2,3,...N$
5	average degree	It is represented by $A = \dfrac{1}{N}\displaystyle\sum_{i=1}^{N}$ Degree (i) for a given network.
6	Degree distribution	It is the probability that a randomly chosen node of the network has degree d, that is, P_d is a probability and must be normalized $\left(P_d = 1\right)$.

- **Regular Network**

  Regular networks [12–14] follow the ordered structure, that is, every node of the system has a same number of links/degree. For degree distribution, no need to use any statistical calculation as the degree of every node is the same. Figure 4.1 is the structure of a regular network where nodes are connected with closed neighbors.

- **Random Network**

  Another structure that usually presents in social get-togethers and network models is a random network [12–14] where any node connects with other nodes of the network. Generally, two models are there to generate

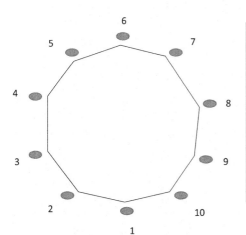

Following are the calculations of the given network:

$(Number\ of\ nodes)\ N = 10$

$(Total\ number\ of\ links)L = \dfrac{N \times degree}{2}$

$= \dfrac{10 \times 2}{2} = 10\ Links$

As it is a regular network, the degree of every node is the same i.e. 2. Hence, no need to calculate degree distribution for the whole system.

FIGURE 4.1   Regular network degree and links.

random networks: (a) n nodes randomly connected with l random links; and (b) each node of network connect based on probability p. If probability in-between processors are greater than or equal to predefined probability, then those nodes can be connected else check it for other nodes. Here, based on probability, we have created a random network as the total number of links used very rarely in the processors' network model. Figure 4.2 represents the statistical calculations of a random network of 10 nodes with degree distribution (Figure 4.3):

- **Scale-free Network**

    A scale-free network [12–14] is nothing but a random network with power-law distribution. In this type of connectivity, arrived new node, pref-erably connected to the highly connected node (hub); therefore, it is also called preferential attachment. In this case, resource availability can be the parameter to choose the node to be connected. In the case of choosing

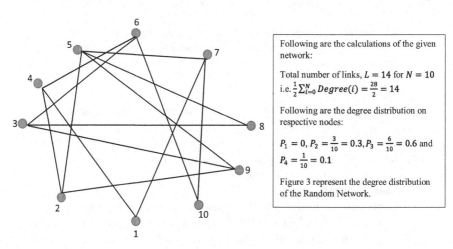

Following are the calculations of the given network:

Total number of links, $L = 14$ for $N = 10$
i.e. $\frac{1}{2}\sum_{i=0}^{N} Degree(i) = \frac{28}{2} = 14$

Following are the degree distribution on respective nodes:

$P_1 = 0, P_2 = \frac{3}{10} = 0.3, P_3 = \frac{6}{10} = 0.6$ and
$P_4 = \frac{1}{10} = 0.1$

Figure 3 represent the degree distribution of the Random Network.

FIGURE 4.2   Random network degree distribution and links.

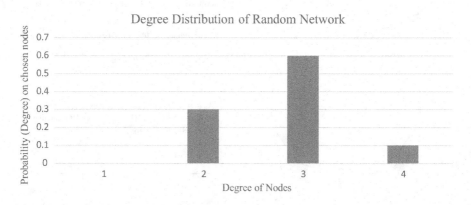

FIGURE 4.3   The random network follows the Poisson degree distribution.

processors for the execution of overloaded tasks, this preferentially attached node will work efficiently. The reason behind its efficiency is connectivity with many processors. Connectivity means this hub has details of all connected nodes that help the source processor to select the target processor easily. The degree distribution of a scale-free network follows power-law distribution [15]. Figure 4.4 represents the physical structure with statistical calculations of a scale-free network, and Figure 4.5 shows the power-law behavior of the said network.

Now we will discuss some basic real-time scheduling algorithms that will be used for the scheduling of real-time tasks on nodes of a mentioned network.

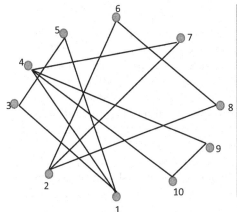

Following are the calculations of the given network:

Total number of links, $L = 12$ for $N = 10$

i.e. $\frac{1}{2}\sum_{i=0}^{N} Degree(i) = \frac{24}{2} = 12$

Following are the degree distribution on respective nodes:

$P_1 = 0, P_2 = \frac{7}{10} = 0.7, P_3 = \frac{2}{10} = 0.2$ and

$P_4 = \frac{1}{10} = 0.1$

Figure 5 represent the degree distribution of a Scale-free network

FIGURE 4.4 Scale-free network degree distribution and links.

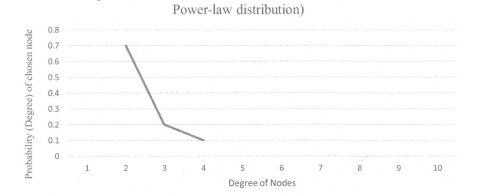

FIGURE 4.5 Power-law distribution of scale-free network.

TABLE 4.2
Real-Time Scheduling Algorithms

Static Priority Scheduling Algorithms	Dynamic Priority Scheduling Algorithms
1. Once the priority is assigned to tasks, it will remain the same throughout execution.	1. Priority of tasks can be changed if some high-priority tasks arrive in the system.
2. Rate monotonic (RM) is a static algorithm that assigns priority based on the shortest period, that is, shorter the period higher the priority.	2. Earliest deadline first (EDF) schedules the tasks having a near deadline. Nearer the deadline higher the priority is the criteria to assign the priority. As the nearest deadline can vary with time, so, it comes under the dynamic category.

- **Real-Time Systems and Scheduling Algorithms**

    A system that deals with tasks based on their temporal behavior is known as RTS [3,4,7,8,16]. RTSs are of two types: soft and hard. In soft RTS, if a single task is unable to meet the deadline, then it affects the performance of the entire system. On the other side, this missing deadline can be the reason for some catastrophic results in hard RTS. To schedule, such tasks scheduling algorithms are required. These scheduling algorithms are of two types: static and dynamic priority-based algorithms. Table 4.2 summarizes the features of both types of algorithms.

    Rate monotonic (RM) and earliest deadline first (EDF) scheduling algorithms arc uscd hcrc for thc cxccution of real-time tasks on a distributed system having nature-inspired connectivity. The next section will explain the proposed architecture of our work.

    To the best of our knowledge, researchers/academicians' more focus is on improvements of real-time scheduling algorithms but not on the type of connectivity, that makes RTS a complex system. This paper aims to check the dependency of tasks execution on the type of physical connectivity of processors.

4.3   PROPOSED ARCHITECTURE

Figure 4.6 demonstrates the complete functioning of the proposed work where real-time scheduling algorithms will implement on all nature-based network connectivity of processors.

The proposed model has following four partitions that are dependent on each other:

   I. All periodic tasks will arrive in **the global task queue**. Through this queue, tasks will get assigned to a local queue of processors of a given distributed system (respective network connectivity).
   II. To schedule tasks on a chosen processor, a real-time scheduling algorithm (EDF/RMS) will be used.
   III. In case of overloading, the victim task can migrate to other connected processors within a network.

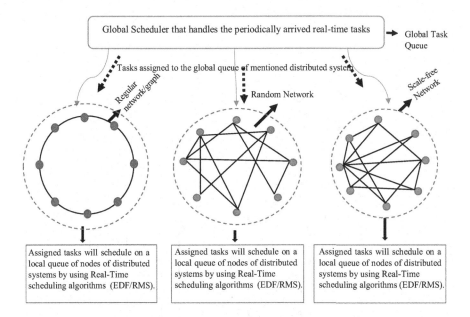

FIGURE 4.6   Proposed architecture.

IV. For migration, with the help of either breadth-first search (BFS) or depth-
first search (DFS), target processor will search for the assignment of the
victim task. Mostly, BFS will be used in all types of networks, because it is
suitable for searching closure nodes.

4.4   PERFORMANCE EVALUATION OF PROPOSED WORK

This section explains the performance of the proposed model with the help of
mathematical equations. Let us say, S is the system of 10 set of processors, that is,
$\{P_1, P_2, P_3, P_4, P_5, P_6, P_7, P_8, P_9, P_{10}\}$. This set of processors can connect in various ways.
Here, we are considering only 3 connectivity structures: regular, random, and scale-
free. $S(P,C)$ is a tuple that represents the network (graph) where S represents system,
P represents processors, and C symbolizes connectivity/link between processors P.
It is already discussed that average degree (A_d) and degree distribution (P(degree))
tell the type of structure system/network/graph is following. To calculate the A_d,
Table 4.3 shows the matrix representation of all above-mentioned graphs/distributed
systems. This matrix helps to check the connectivity in-between nodes (degree of
a node) during implementation that further tells the type of physical structure the
system is following.

With the help of following theorem, the effect of processors connectivity on real-
time task execution can be explained.

Theorem I: Real-Time tasks' timely execution is dependent on the type of processors connectivity in distributed system.

Solution: Concerning Section 4.2, Figures 4.1, 4.2, and 4.4 have shown the different connectivities of processors.

TABLE 4.3

Matrix Representation of Graph to Calculate the Average Degree and Check the Connectivity between Nodes

S. No.	Distributed Structure	Matrix Representation [17]											Average Degree (A_d)
1	Regular	p	1	2	3	4	5	6	7	8	9	10	Total Degree (D_p)	For undirected
		1	0	1	0	0	0	0	0	0	0	1	2	$A_d = \dfrac{1}{N}\displaystyle\sum_{p=1}^{10} D_p = \dfrac{1}{10}\times 20 = 2$
		2	1	0	1	0	0	0	0	0	0	0	2	
		3	0	1	0	1	0	0	0	0	0	0	2	
		4	0	0	1	0	1	0	0	0	0	0	2	
		5	0	0	0	1	0	1	0	0	0	0	2	
		6	0	0	0	0	1	0	1	0	0	0	2	
		7	0	0	0	0	0	1	0	1	0	0	2	
		8	0	0	0	0	0	0	1	0	1	0	2	
		9	0	0	0	0	0	0	0	1	0	1	2	
		10	1	0	0	0	0	0	0	0	1	0	2	
		Total Degree (D_p)	2	2	2	2	2	2	2	2	2	2	$\displaystyle\sum_{p=1}^{10} D_p = 20$	

(Continued)

TABLE 4.3 (Continued)

Matrix Representation of Graph to Calculate the Average Degree and Check the Connectivity between Nodes

S. No.	Distributed Structure	Matrix Representation [17]											Average Degree (A_d)
2	Random	p	1	2	3	4	5	6	7	8	9	10	Total degree (D_p)	$A_d = \dfrac{1}{N}\displaystyle\sum_{p=1}^{10} D_p = \dfrac{1}{10}\times 28 = 2.8 = 2.8 \cong 3$
		1	0	0	0	1	0	0	1	0	0	0	2	
		2	0	0	0	1	1	0	0	0	1	0	3	
		3	0	0	0	0	0	1	0	1	1	0	3	
		4	1	1	0	0	0	1	0	0	0	0	3	
		5	0	1	0	0	0	0	1	1	1	0	4	
		6	0	0	1	1	0	0	0	0	0	1	3	
		7	1	0	0	0	1	0	0	0	0	1	3	
		8	0	0	1	0	1	0	0	0	0	0	2	
		9	0	1	1	0	1	0	0	0	0	0	3	
		10	0	0	0	0	0	1	1	0	0	0	2	
		Total degree (D_p)	2	3	3	3	4	3	3	2	3	2	$\displaystyle\sum_{p=1}^{10} D_p = 28$	

(Continued)

TABLE 4.3 (Continued)

Matrix Representation of Graph to Calculate the Average Degree and Check the Connectivity between Nodes

S. No.	Distributed Structure	Matrix Representation [17]											Average Degree (A_d)
3	Scale-free	p	1	2	3	4	5	6	7	8	9	10	Total degree (D_p)
		1	0	0	1	1	1	0	0	0	0	0	3
		2	0	0	0	0	0	1	1	1	0	0	3
		3	1	0	0	0	1	0	0	0	0	0	2
		4	1	0	0	0	0	0	1	0	1	1	4
		5	1	0	1	0	0	0	0	0	0	0	2
		6	0	1	0	0	0	0	0	1	0	0	2
		7	0	1	0	1	0	0	0	0	0	0	2
		8	0	1	0	0	0	1	0	0	0	0	2
		9	0	0	0	1	0	0	0	0	0	1	2
		10	0	0	0	1	0	0	0	0	1	0	2
		Total degree (D_p)	3	3	2	4	2	2	2	2	2	2	$\sum\limits_{p=1}^{10} D_p = 24$	$A_d = \dfrac{1}{N}\sum\limits_{p=1}^{10} D_p = \dfrac{1}{10} \times 24 = 2.4 \cong 2$

TABLE 4.4

Equations Explain the Working of the Real-Time Scheduling Algorithm (EDF & RMS)

Eq. No.	Equation	Description
4.1	$\{t_1, t_2, t_3, t_4, \ldots\ldots\ldots, t_n\} \in Q$	n periodic tasks arrive in the global queue Q of system S
4.2	$\{t_1, t_2, t_3, t_7\} \in P_1,$ $\{t_4, t_5, t_8, t_{10}\} \in P_2, \{t_6, t_{11}, t_{13}, t_{17}\} \in P_3,$ $\{t_9, t_{12}, t_{14}, t_{18}\} \in P_4, \ldots\ldots\ldots,$ $\{t_n, t_{n+2}, t_{n+3}, t_{n+k}\} \in P_{10}.$	Q is a priority queue that assigns tasks to randomly chosen processors on an FCFS basis. If t_1 comes first in the Q then it will assign to the randomly chosen processor of the system first, then t_2, t_3 and so on.
4.3	$\sum_{n=1}^{k} t_n u = P_i U \leq 1$ and $t_n u = \dfrac{t_n e}{t_n \text{ Period}}$	$P_i U$ is the total utilization of ith processor, and $t_n u$ is the utilization of nth task. As per the EDF scheduling algorithm $P_i U \leq 1$ $t_n e$: worst-case execution time of nth task and t_n Period : an inter-arrival period of nth task.
4.4	$\sum_{n=1}^{k} t_n u = P_i U \leq n\left(2^{\frac{1}{n}} - 1\right)$ and $t_n u = \dfrac{t_n e}{t_n \text{ Period}}$	In RMS scheduling algorithm, $P_i U \leq n\left(2^{\frac{1}{n}} - 1\right),$

First, explain the common picture of tasks appearance (refer to Table 4.4) in a global task queue (Q) with arrival time $(t_n a)$, execution time $(t_n e)$, inter-arrival period $(t_n \text{ Period})$, and deadline $(t_n \text{ Deadline})$.

Table 4.5 explains the working of EDF and RMS algorithm in overloading case of **regular network** (Figure 4.1)

Table 4.6 explains the working of the EDF and RMS algorithm in overloading case of **the random network** (Figure 4.2)

A scale-free network is also called a preferentially connected model where the node with more resources becomes the "highly connected" node of the network. Its ratio with the remaining nodes of network remains constant with the expansion or scalability of a given work. With the help of Equation 8, the degree of nodes of network is calculated that will tell the highly connected nodes of the network, that is, nodes with highest degree become the hub (very connected node) of the network. Although scale-free is a type of random network with the following feature:

- It can be used to find the processor for real-time tasks efficiently due to the connectivities of a hub with miscellaneous performed processors.
- For task migration purposes, very easily less-utilized processors can be targeted due to heterogeneity and high connectivity with other processors in a system.
- Its scale-up behavior provides more resources after the addition of a new node in the system.

TABLE 4.5

Equations Explain the Working of Regular Network in a RTSs Scenario

Eq. No.	Equation	Description
4.5	$D_{P_i} = 2$ where i varies from 1 to 10	The degree of each processor is same, that is, 2, as all processors connect with nearby neighbors (refer matrix representation).
4.6	$iff \left(P_iU = \sum_{n=1}^{k} t_n u > 1 \right))$	If P_1 utilization exceeds the maximum utilization, that is, 1, then the overloaded task needs to migrate toward connected processors of P_1, that is, P_{10} or P_2.

4.7	$P_i\left[D_{P_i} \right] = \left[P_{i+1} \right]$
	 $iff\ \left(P_{i+1}U < 1 \right)$
	 {
	 t_n of $P_i \rightarrow \min_{i \rightarrow 1\ to\ 9} P_{i+1}U$ where i varies from 1 to 2
	 }
		As per matrix representation of regular network (Table 4.3), Degree(P_1) is 2, which means only two processors are connected with P_1. So, a processor with the least utilization among 2 processors will be chosen for scheduling of overloaded task t_n of P_1.

4.8	$P_iU = \sum_{n=1}^{n-1} t_{n-1} u \leq 1$	After migration of nth task, the utilization of P_1 becomes less than or equal to 1.

BFS algorithm [18–20] is used here to search the processor of minimum utilization among all connected processors.

Overall, a scale-free network connectivity structure can be considered to design an optimal distributed system for the timely execution of real-time tasks.

To evaluate the considerable network connectivity, we consider the following parameters:

I. **Resource availability at a common point:** The meaning of resource availability is the degree of a respective node (one node has been connected with how many nodes). In a regular network, the degree of each node is the same, and they are connected with nearest neighbors; hence, very limited resources are available at a single node for task execution. Similarly, with the help of a matrix, very easily node with maximum resource availability can be searched. In a random network, resource availability is less as compared to a scale-free network. In scale-free network, highly connected node will become maximum resource provider.

II. **Number of migration:** The task that crosses the utilization value of the given processor becomes the overloaded task that can be responsible for the

TABLE 4.6
Equations Explain the Working of Random Network in RTSs Scenario

Eq. No.	Equation	Description
4.7	The random network has been generated according to $G(P, \text{Pr})$	As per Figure 4.2, $P = 10$ processors and probability Pr that are links between any two processors are $\frac{1}{5}$ or 0.20.
4..8	$D_{P_i} = \sum_{j=1}^{10} \text{mat}[i][j]$ Or $D_{P_j} = \sum_{i=1}^{10} \text{mat}[j][i]$	Degree (number of links) of respective processor P_i where i varies from 1 to 10 has been calculated by addition of row-/column-wise entries (refer to Table 3).
4.9	$iff\left(P_i U = \sum_{n=1}^{k} t_n u > 1 \right)$	If P_1 utilization exceeds the maximum utilization, that is, 1, then the overloaded task needs to migrate toward connected processors of P_1.

4.10	$P_i\left[D_{P_i} \right] = [\text{connected processors}]$
	 $iff(P_k U < 1)$
	 {
	 t_n of $P_i \rightarrow \min_k P_k U$ where
	 $k \in$ connected processors of P_i
	 }
		As Degree(P_i) varies here, that is, depends on some links. As per Figure 4.3,
		 $P_1[2] = \{P_4, P_7\}$, $P_2[3] = \{P_4, P_5, P_9\}$, $P_3[3] = \{P_6, P_8, P_9\}$,
		 $P_4[3] = \{P_1, P_2, P_6\}$, $P_5[4] = \{P_2, P_7, P_8, P_9\}$,............,...,
		 $P_{10}[2] = \{P_6, P_7\}$
		 respectively. Processor with least utilization among connected processors (P_k) will be chosen for scheduling of overloaded task t_n of P_1.

4.11	$P_i U = \sum_{n=1}^{n-1} t_{n-1} u \leq 1$	After migration of nth task, the utilization of P_1 becomes less than or equal to 1.

BFS algorithm [18–20] is used here to search the processor of minimum utilization among all connected processors.

poor performance of system. To execute such tasks, migration of overloaded tasks is required that also takes time. Some migration means task migration continues till it finds the least utilized processors. In the case of a scale-free network, this migration time is less due to the presence of a highly connected node.

III. **Quick destination searching approach:** BFS approach will be used to search the least utilization processor for task scheduling.

IV. **Tardiness:** The lateness in task execution is known as tardiness (task completion time − task deadline). It can be increased if migration time increases or less if respective tasks migration time is less.

Table 4.7 concludes the differences between all three structures that are discussed in this paper for timely execution of real-time tasks.

TABLE 4.7

Performance Matrix of Discussed Network Structure Use for Real-Time Tasks Execution

Performance Evaluation Parameters	Regular Network	Random Network	Scale-free Network
Resource availability at a common point	Limited and fixed at each node	Limited but not fixed	Neither it is limited nor fixed here. The addition of nodes is possible.
Number of migration	If connected processors are fully utilized, the number of migrations is more, and it will take time.	As compared to regular, a fewer number of migrations will be here due to the availability of small hubs.	Due to heterogeneous clustering, it's easy to select an available processor that reduces the number of migrations
Quick destination searching approach	BFS	BFS	BFS
Tardiness of migrated tasks	More	More	Less

4.5 CONCLUSION AND FUTURE SCOPE

As proposed, the mathematical model description, system architecture, and synchronous functioning have been successfully described in this paper. This paper will serve as a base for the further complex simulation. We anticipate functional efficiency of various network models of RTDS may follow a general characteristics pattern. This may lead us to summarily conclude that which typical network structure may have higher efficiency with scheduling algorithm. Furthermore, algorithms may be tested upon the proposed scenario, and performance will be evaluated based on tardiness and the number of migrations of overloaded tasks.

REFERENCES

1. Malony, A.D., & Shende, S. (2000). Performance technology for complex parallel and distributed systems. In: Kacsuk, P., & Kotsis, G. (eds) *Distributed and Parallel Systems. The Springer International Series in Engineering and Computer Science*, vol 567. Springer, Boston, MA. https://doi.org/10.1007/978-1-4615-4489-0_5
2. Coulouris, G., Dollimore, J., & Kindberg, T. (2000). Distributed systems: Concepts and design edition 3. *System*, 2(11), 15.
3. Berns, K., Köpper, A., & Schürmann, B. (2021). Real-Time systems. In: Berns, K., Köpper, A., & Schürmann, B. (eds) *Technical Foundations of Embedded Systems* (pp. 339–361). Springer, Cham.
4. Bendib, S. S., Kalla, H., Kalla, S., & Hocine, R. (2021). A self-organized scheduling algorithm for embedded real-time systems. *International Journal of Embedded and Real-Time Communication Systems (IJERTCS)*, 12(2), 57–73.

5. Singh, A., Sharma, S., & Singh, J. (2021). Nature-inspired algorithms for wireless sensor networks: A comprehensive survey. *Computer Science Review, 39,* 100342.

6. Barabási, A. L. (2013). Network science. *Philosophical Transactions of the Royal Society A: Mathematical, Physical and Engineering Sciences, 371*(1987), 20120375.

7. Donga, J., Holia, M. S., & Patel, N. M. The classification and comparative study of real-time task scheduling algorithms based on various parameters. In *Applications of Artificial Intelligence in Engineering: Proceedings of First Global Conference on Artificial Intelligence and Applications (GCAIA 2020)* (p. 239). Springer Nature, Jaipur.

8. Laplante, P. A. (2004). *Real-Time Systems Design and Analysis* (p. xxi). Wiley, New York.

9. Liu, C. L., & Layland, J. W. (1973). Scheduling algorithms for multiprogramming in a hard real-time environment. *Journal of the ACM (JACM), 20*(1), 46–61.

10. Bao, X., & Li, J. (2005). Matching code-on-graph with network-on-graph: Adaptive network coding for wireless relay networks. In *Proceedings of Allerton Conference on Communication, Control and Computing,* Illinois.

11. Gross, J. L., Yellen, J., & Anderson, M. (2018). *Graph Theory and Its Applications.* Chapman and Hall/CRC, New York.

12. Cowan, R., & Jonard, N. (2004). Network structure and the diffusion of knowledge. *Journal of Economic Dynamics and Control, 28*(8), 1557–1575.

13. Bansal, S., Khandelwal, S., & Meyers, L. A. (2009). Exploring biological network structure with clustered random networks. *BMC Bioinformatics, 10*(1), 1–15.

14. Pulipati, S., Somula, R., & Parvathala, B. R. (2021). Nature-inspired link prediction and community detection algorithms for social networks: A survey. *International Journal of System Assurance Engineering and Management, 12,* 1–18.

15. Kak, S. (2021). The base-e representation of numbers and the power law. *Circuits, Systems, and Signal Processing, 40*(1), 490–500.

16. Kumar, H., & Tyagi, I. (2021). A hybrid model for tasks scheduling in a distributed real-time system. *Journal of Ambient Intelligence and Humanized Computing, 12*(2), 2881–2903.

17. Hu, D., Broersma, H., Hou, J., & Zhang, S. (2021). On the spectra of general random mixed graphs. *The Electronic Journal of Combinatorics, 28*(1), P1–3.

18. Lyu, Q., & Gong, B. (2021). Speedup breadth-first search by graph ordering. *International Journal of Computer and Information Engineering, 15*(7), 428–435.

19. Kaur, A., Sharma, P., & Verma, A. (2014). A appraisal paper on Breadth-first search, Depth-first search and Red black tree. *International Journal of Scientific and Research Publications, 4*(3), 2–4.

20. Beisegel, J., Denkert, C., Köhler, E., Krnc, M., Pivač, N., Scheffler, R., & Strehler, M. (2021). The recognition problem of graph search trees. *SIAM Journal on Discrete Mathematics, 35*(2), 1418–1446.

5 Memory Characterization of GP-GPU Applications

Govind Sreekar Shenoy
Nitte Meenakshi Institute of Technology

CONTENTS

5.1 INTRODUCTION

Computer architecture is going through an era of revolution. Multi-cores have become common and are pervasive over the last fifteen years. Graphics processors have also become prominent over the last 15 years. Graphics processors typically execute graphics applications. Graphics applications are streaming in nature with little or no locality. So typically graphics processors do not use caches to accelerate the applications since they are bandwidth-sensitive.

General-purpose applications for GPUs (GP-GPUs) have become prominent. In GP-GPUs, the graphics processor typically executes general-purpose applications, hence the name GP-GPU. General-purpose applications are latency-sensitive. Traditional graphics applications typically exhibit no locality; hence, caches for GPUs have been relatively less unstudied. With the advent of GP-GPUs over the last 10–15 years, caches have become critical. The main issue in a cache for GPU is the limited silicon area for the cache. Typically, the GPU is provisioned for exploiting parallelism with a huge number of cores and thousands of threads. Thus, the space for caches is very minimal in a GPU.

There have been studies [1] that optimize the L1 data cache of the GPU by limiting the number of threads. There have also been works that look at prioritizing the access pattern of a cache [2]. There have also been works that investigate tightly coupling the thread scheduling mechanism with the cache management algorithms [3].

DOI: 10.1201/9781003320333-5

There have also been numerous works [4–6] that leverage cache behavior in order to improve the performance of the L1 data cache.

In this work, we analyze the performance of the L1 data cache of the GPU, in terms of the miss rate. We vary the L1 data cache size from 32 to 256 KB and use intermediate values of 64 and 128 KB. We further vary the associativity of the cache from 16 to 256 with intermediate values of 32, 64, and 128. We also vary the number of banks of the L1 data cache with values of 1, 2, 4, and 8. We observe that the L1 data cache miss rate is very high, sometimes even 100%. This is due to the huge working set size of the GP-GPU applications. The huge working set size of the L1 data cache limits the performance, and we do not observe significant difference in the L1 data cache miss rate as we vary the cache size. On varying the associativity, we observe there is little difference in the miss rates. This is due to significant cache thrashing which results in huge miss rate.

This chapter is organized as follows: Section 5.2 provides a background on GPUs and introduces GP-GPUSim and the Rodinia benchmark. Section 5.3 surveys the literature and discusses all the related work. Section 5.4 discusses the simulation methodology used in this paper. Section 5.5 introduces and analyzes the results obtained. Finally, Section 5.6 concludes this work and discusses the future directions.

5.2 BACKGROUND

5.2.1 GPU ARCHITECTURE

Figure 5.1 shows the micro-architecture of a conventional GPU processor. The architecture consists of multiple cores often referred to in literature as symmetric multi-processors (SMs). Each SM contains its own private L1 data cache and executes in a

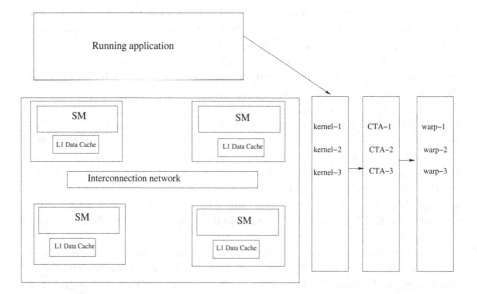

FIGURE 5.1 GPU micro-architecture.

Single Instruction Multiple Thread (SIMT) fashion. In SIMT execution, each thread executes an instruction and multiple such threads execute in a lock step fashion. A bunch of threads often referred to as a warp execute together in a lock step fashion. In case there is a branch instruction, then a subset of threads in the warp executes one part of the control flow while the remaining threads execute the other part of the control flow. NVIDIA GPU processors commonly use a bunch of 32 threads to constitute a warp. A group of warps constitute a cooperative thread array (CTA) (shown in Figure 5.1). Furthermore, a group of CTAs constitute a kernel, and there are multiple such kernels which are spawned by the host processor and scheduled on the GPU as shown in Figure 5.1. Thus, the GPU execution can be viewed as a hierarchy of steps, wherein at the lowest level, a SM executes a bunch of 32 threads in a SIMT fashion. This bunch of threads referred to as warps is the next level in the thread execution hierarchy, which forms a CTA. Furthermore, a group of CTA forms a kernel. Another feature of GPUs is that the size of the register file is greater than the size of the private L1 data cache. Each SM in a NVIDIA Fermi had 32 KB of registers while the L1 data cache size is 16 KB.

A SM essentially consists of a L1 data cache which is private to each SM. Furthermore, a SM also contains register files and scratch pad memory. There is a shared L2 cache which is shared by all the SMs. The shared L2 cache communicates with the L1 data cache using an interconnection network. The L2 cache interfaces with the DRAM which is further down in the memory hierarchy and fetches data from the DRAM. A GPU has its own private DRAM which is separate from the DRAM of the CPU. Before execution of a kernel, the CPU transfers data from the CPU to the GPU DRAM. Once the data transfer is complete, the kernel starts executing.

There are multiple SMs with the NVIDIA Fermi which is one of the earliest GPUs. It had 16 SMs, thus resulting in $16 \times 32 = 512$ threads. Modern NVIDIA GPUs like NVIDIA Ampere contain 108 SMs, thus resulting in a massively parallel processor.

The L1 data cache is, as mentioned earlier, a private data cache which is private to each SM. The L1 data cache in Fermi can be configured dynamically to either 32 or 48 KB wherein the L1 data cache shares silicon area with the scratch pad. When the L1 data cache sized is increased dynamically, it results in the scratchpad memory being decreased accordingly.

5.2.2   GPGPUSIM SIMULATOR

In this chapter, we use the very popular simulator from University of British Columbia named GPGPUSim [7]. GPGPUSim was developed in 2009 and has recently undergone major enhancements. The initial version of GPGPUSim supported CUDA version 4.0, but with the latest release, it supports CUDA version 11. We have varied the L1 data cache parameters in GPGPUSim and reported the L1 data cache miss rate.

5.2.3   CACHE CHARACTERIZATION

Cache performance is critical for CPU applications. With the advent of general-purpose GPU applications, cache performance has become a bottleneck in GPUs.

TABLE 5.1
Rodinia Benchmark Suite

Application	Domain
K-means	Data mining
Needleman–Wunsch	Bioinformatics
Hotspot	Physics simulation
Back propagation	Pattern recognition
SRAD	Image processing
Leukocyte tracking	Medical imaging
BFS	Graph algorithms
Stream cluster	Data mining
B+ tree	Graph algorithms
Gaussian elimination	Dense linear algebra
Heart wall	Structured grid
Hotspot	Structured grid
Hybrid sort	Sorting
LU decomposition	Dense linear algebra

Caches primarily exploit temporal locality and spatial locality. When the same line is being reused over and over again, it is referred to as temporal locality. Thus, applications typically tend to reuse memory resulting in temporal locality exploitation. Cache lines which are adjacent also result in access of adjacent cache lines, and this is referred to as spatial locality. Typically, caches use blocks of data to access adjacent memory locations, thus resulting in spatial locality exploitation.

In this work, we vary the cache size, cache associativity, and the number of cache banks. We conduct a thorough design space exploration of these cache parameters in order to better characterize the locality of GP-GPU applications.

5.2.4 BENCHMARKS

We use the Rodinia benchmark suite [8] in our evaluation. Table 5.1 lists the benchmarks in the Rodinia suite. We use these benchmarks in our evaluation as Rodinia is one of the most widely used benchmarks in GP-GPU study. The Rodinia benchmarks are vast and varied, and range from Physics to web mining.

5.3 RELATED WORK

L1 data caches for graphics processors have gained traction over the last few years in literature. One of the earliest work exploring data caches for graphics processors was by Rogers et al. [1] who designed a cache specifically tailored for graphics processor computation. They tailored the scheduling mechanisms of warps of threads in a manner that increases the locality of L1 data cache. This is one of the first works that studies the impact of scheduling on the L1 data cache performance.

Jia et al. [9] characterize the performance of GPUs with caches. They further provide a taxonomy for reasoning about different types of access patterns and locality. They further present an algorithm that is automated and applied at compile time in order to identify an application's memory access pattern. They observe important performance benefits in using their static scheme for a wide range of applications.

Jia et al. [2] explore prioritizing methods to improve cache performance. They explore two prioritization methods, namely request reordering and cache bypassing. They name their scheme as memory request prioritization buffer (MRPB). MRPB releases requests into the cache in a more cache friendly manner. They observe significant reduction in cache contention with their proposed scheme.

Li et al. [3] propose to tightly couple the thread scheduling mechanism with the cache management algorithms. They propose a mechanism such that the GPU L1 data cache pollution is minimized while the off-chip memory throughput is enhanced. They propose a priority-based cache allocation that provides preferential cache capacity to a subset of high-priority threads while allowing lower priority threads to execute with contention.

Xie et al. [6] propose coordinated static and dynamic cache bypassing. At compile time, they identify loads which have a behavior of either caching or bypassing. They use profiling to identify such loads. In the dynamic bypassing technique, at run time, they classify a load as cacheable or bypassable and mark the load accordingly and either cache it or bypass it. They observe important performance improvement in using their scheme.

Ausavarungnirun et al. [4] observe that GPU warps exhibit heterogeneous memory divergence behavior at the shared cache. As a result, few warps result in a cache hit while other warps result in a cache miss. They further observe that warps exhibit divergence wherein few threads in a warp take one control path while the remaining threads take a different control path. They also observe that memory requests going to the shared cache incur high queuing delays. As a consequence, they propose a set of techniques to mitigate the memory divergence and queuing delays in memory accesses.

Ibrahim et al. [5] observe that there is a very high sharing of data between the L1 data caches across the cores. So between two cores in a GPU, cache lines are getting reused. This results in wastage of L1 data cache capacity. In order to mitigate sharing of cache lines, they propose a decoupled L1 data cache in such a manner that the L1 data cache is separate from the GPU core. This innovative cache design reduces the data replication across the L1 data caches and improves the bandwidth utilization. They perform a comprehensive evaluation and observe important performance benefits in their decoupled L1 data cache design.

5.4 SIMULATION METHODOLOGY

We have done our memory characterization using the GPGPUSim [7] which is a cycle-accurate simulator. The GPGPUSim has already been explained in Section 2.2. We use the Rodinia benchmarks which have also been discussed in Section 2.4. We vary the cache size, cache associativity, and the number of banks as shown in Table 5.2.

TABLE 5.2
Cache Space Exploration

Cache Size	Cache Associativity	Number of Banks
16 KB	256	1
32 KB	128	2
64 KB	64	4
128 KB	32	8
256 KB	16	

TABLE 5.3
NVIDIA Titan V Configuration

Processor Parameter	Value
SMs	80
SM configuration	Warps per SM = 64, scheduler per SM = 4, warps per sched = 16
Exec units	4 SPs, 4 SFUs
Memory unit	Fermi coalescer (32 thread coalescer)
L1 cache	32 KB, 128 B line size, 4 way, write evict
L2 cache	4.5 MB, 24 Banks, 128 B line size, 32 ways
Interconnection network	40x24 crossbar, 32 B flit size
DRAM model	GDDR5 model, BW = 652 GB/second

We have used the NVIDIA TITAN-V GPU processor configuration [10]. The GPU configuration is shown in Table 5.3. TITAN is a 2017 GPU processor and is one of the most recent GPU processors. We have varied the L1 data cache of the TITAN.

5.5   RESULTS

We have varied the L1 data cache from 32 to 256 KB with intermediate values of 64 and 128 KB. Further, we have varied the associativity from 32 to 256 with values of 64, 128. Moreover, we have also varied the number of banks of the L1 data cache with values.

Figure 5.2 shows the results of varying the cache size from 32 to 256 KB. We see that the benchmarks in general are invariant to the cache size. This implies that the GP-GPU applications in general are cache-insensitive. This is also expected with the very high miss rate of GP-GPU applications. We notice that most of the benchmarks have a very high miss rate implying that the working set size of these benchmarks is huge and cannot be contained in the L1 data cache. For benchmarks like breadth-first search (BFS), B+tree, and CFD, there is a marginal variation with the miss rate dropping down as we increase the cache size. This behavior is also seen in SRAD and stream cluster.

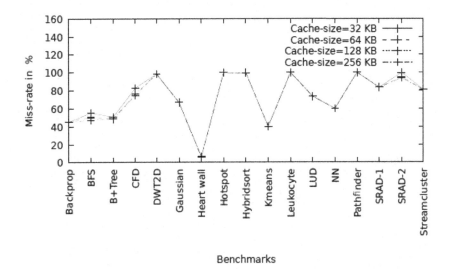

FIGURE 5.2   Varying the cache size.

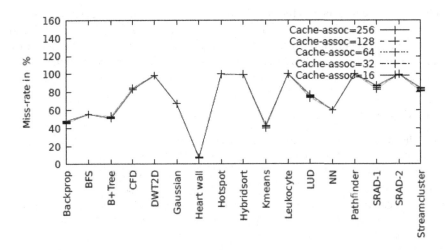

FIGURE 5.3   Varying the cache associativity.

Figure 5.3 shows the results of varying the cache associativity from 16 to 256. We observe that the benchmarks have a constant miss rate irrespective of the cache associativity. This implies that there is significant cache thrashing happening in GP-GPU benchmarks which is also substantiated by the high miss rate seen in these benchmarks. Benchmarks like B+tree, Computational fluids dynamics (CFD), and Speckle Reducing Anisotropic Diffusion (SRAD)show a slight variation with increasing associativity in the miss rate, but it is not significant.

Figure 5.4 plots the variation of the number of cache banks in the L1 data cache by varying it from 1 to 8 with intermediate values of 2 and 4. We again see that the

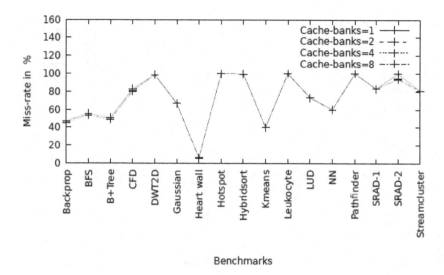

Benchmarks

FIGURE 5.4   Varying the number of cache banks.

benchmarks are invariant to the number of cache banks with a marginal decrease in cache miss rate as we increase the number of banks for Backprop, B+tree, BFS, and CFD. But, in general, the benchmarks are invariant to the number of cache banks. This is again expected as the miss rate for these benchmarks is very high; thus, there is little benefit of caches as the working set size is very huge.

5.6   CONCLUSION

This chapter studied the memory characterization of general-purpose graphics processing unit (GP-GPU) applications. We have used the latest GPU available and done a simulation study. In the study, we varied the L1 data cache size from 32 to 256 KB with intermediate values of 64 and 128 KB. We also varied the cache associativity of L1 data cache from 16 to 256 with intermediate values of 32, 64, and 128. We further varied the number of cache banks of L1 data cache from 1 to 8 with values of 2 and 4 in between.

We observe that the cache size, associativity, and the number of banks have little impact on the miss rate. This is primarily because the miss rate is very high for most of the applications, thus implying a very huge working set size which cannot be contained despite having cache size of 256 KB. The high miss rate limits the impact of cache size and number of cache banks. We also observe that there is little impact on increasing the associativity due to the high cache thrashing that is happening in the L1 data cache. Thus, we can conclude that GP-GPU applications are cache-insensitive.

We have studied the memory characterization of GP-GPU applications. A similar study could be done for machine-learning applications, and it will be interesting to observe the impact of varying the L1 data cache and analyze the performance of machine-learning applications. A further area of study could be to incorporate L1 data cache techniques used in CPUs to GPUs and analyze the performance of GPUs.

REFERENCES

1. Rogers, T.G., OConnor, M., Aamodt, T.M.: Cache-conscious wavefront scheduling. In: *2012 45th Annual IEEE/ACM International Symposium on Microarchitecture*. IEEE (Dec 2012). https://doi.org/10.1109/micro.2012.16.
2. Jia, W., Shaw, K.A., Martonosi, M.: MRPB: Memory request prioritization for massively parallel processors. In: *2014 IEEE 20th International Symposium on High Performance Computer Architecture (HPCA)*. IEEE (Feb 2014). https://doi.org/10.1109/hpca.2014.6835938.
3. Li, D., Rhu, M., Johnson, D.R., O'Connor, M., Erez, M., Burger, D., Fussell, D.S., Redder, S.W.: Priority-based cache allocation in throughput processors. In: *2015 IEEE 21st International Symposium on High Performance Computer Architecture (HPCA)*. IEEE (Feb 2015). https://doi.org/10.1109/hpca.2015.7056024.
4. Ausavarungnirun, R., Ghose, S., Kayiran, O., Loh, G.H., Das, C.R., Kan- demir, M.T., Mutlu, O.: Exploiting inter-warp heterogeneity to improve GPGPU performance. In: *2015 International Conference on Parallel Architecture and Compilation (PACT)*. IEEE (Oct 2015). https://doi.org/10.1109/pact.2015.38.
5. Ibrahim, M.A., Kayiran, O., Eckert, Y., Loh, G.H., Jog, A.: Analyzing and leveraging decoupled l1 caches in GPUs. In: *2021 IEEE International Symposium on High-Performance Computer Architecture (HPCA)*. IEEE (Feb 2021). https://doi.org/10.1109/hpca51647.2021.00047.
6. Xie, X., Liang, Y., Wang, Y., Sun, G., Wang, T.: Coordinated static and dynamic cache bypassing for GPUs. In: *2015 IEEE 21st International Symposium on High Performance Computer Architecture (HPCA)*. IEEE (Feb 2015). https://doi.org/10.1109/hpca.2015.7056023
7. Bakhoda, A., Yuan, G.L., Fung, W.W.L., Wong, H., Aamodt, T.M.: Analyzing CUDA workloads using a detailed GPU simulator. In: *2009 IEEE International Symposium on Performance Analysis of Systems and Software*. IEEE (Apr 2009). https://doi.org/10.1109/ispass.2009.4919648.
8. Che, S., Boyer, M., Meng, J., Tarjan, D., Sheaffer, J.W., Lee, S.H., Rodinia, S.K.: A benchmark suite for heterogeneous computing. In: *2009 IEEE International Symposium on Workload Characterization (IISWC)*. IEEE (Oct 2009). https://doi.org/10.1109/iiswc.2009.5306797.
9. Jia, W., Shaw, K.A., Martonosi, M.: Characterizing and improving the use of demand-fetched caches in GPUs. In: *Proceedings of the 26th ACM International Conference on Supercomputing - ICS 12*. ACM Press (2012). https://doi.org/10.1145/2304576.2304582.
10. Titan-V, N.: NVIDIA Titan-V GPU architecture. https://www.nvidia.com/en- us/titan/titan-v/ (2017), [Online; accessed 28-August-2021]

6 A Fuzzy Logic-Based Emergency Vehicle Routing Technique for Vehicular Ad-hoc Networks

Rashmi Ranjita and Sasmita Acharya
Veer Surendra Sai University of Technology

CONTENTS

6.1 INTRODUCTION

Vehicular ad-hoc network (VANET) is a wireless network of moving objects called nodes. Nowadays, vehicles are equipped with computing technologies and wireless communication devices that make it a challengeable research area in intervehicular communication and vehicle-to-infrastructure communication. VANETs have numerous applications like collision prevention, safe drive, crossing blind curves, dynamic route scheduling, real-time traffic monitoring, etc. One important application of VANET is providing Internet to moving vehicles. Currently, most advanced vehicles are provided with intelligent communication system. Various sensors make it possible; they gather the surrounding information and transmit it to base station which is then transmitted to network from where it becomes useful data. All these communications occur wirelessly. The sensors that facilitate wireless communication with computing devices in vehicles play a vital role in public healthcare.

In urban areas, traditional wireless sensor networks (WSNs) perform data collection for various decision-making systems like traffic management, public healthcare service, environment monitoring, etc. The paper proposes a design for IoT system that monitors temporary road congestion or blockage in smart cities. Different methods are there to collect traffic-related data [1]. Floating car data (FCD) is one of the methods for collecting traffic data. FCD collects traffic data from vehicles having provision of on-road sensors and global positioning system (GPS) unit. FCD also has a computing device that finds a route to the nearest health center. There are many source routing engines like pgRouting, open trip planner, "open-source routing machine (OSRM)" that build a routing graph for finding the shortest path to the destination. OSRM collects live traffic updates and uses multilevel Dijkstra algorithm to find the shortest routing graph from the current location to the nearest healthcare center. In this paper, the input parameters for the proposed fuzzy inference system (FIS) are average vehicle speed on road, sound level, CO_2 level, CO level, temperature, and radio frequency level at a particular location. Congestion level is the output of the FIS. From this congestion level, the source routing engines build the routing graph that assists the emergency vehicles.

"The rest of the paper is organized as mentioned later". The work done by different researchers in this area is discussed as related work in Section 6.2. Section 6.3 discusses the proposed FIS to detect congestion on road and IoT-based "emergency vehicle routing system". Section 6.4 presents the simulation results of the proposed system. Section 6.5 concludes the paper.

6.2   RELATED WORK

The persons who are injured in road accidents, affected in natural disasters, and injured in fire accidents need emergency healthcare services, alert for road blockage and emergency vehicle routing. There are no major studies conducted related to IoT-based emergency vehicle routing according to [2]. The paper [3] designed a traffic view framework that gathers and disseminates information about vehicles on road. It takes static view of traffic status on the road and gives alert messages to follow the appropriate route. It does not consider the traffic data that are received from on-road sensors. CrowdITS is a crowd-sourcing routing technique described in Ref. [4]. According to CrowdITS technique, smartphone users provide on-road traffic information with some monetary conditions. The CrowdITS server processes and aggregates these collected information. Cloud-to-device messaging service of Google, delivers the pulled information about the on-road traffic to the users of crowd base on their geo-location. The paper [5] proposed a fuzzy logic-based accident monitoring system that consists of tilt sensors, sound sensors, and Raspberry Pi, and is able to detect the road accidents. A traffic flow prediction model is presented in Ref. [6] which proposed a solution to emergency vehicle routing problem.

The authors of Ref. [7] have proposed a model for post-disaster emergency vehicle such as ambulance dispatching and routing simulation. The input parameters for the model are hospital location, on-road traffic condition, etc. It computes the route for the ambulance to deliver the patient efficiently. The paper [8] proposes an intelligent location-based data aggregation technique for VANETs considering the real-time traffic congestion estimation. It uses "Kalman filter data fusion technique" which finds the congestion level. According to Ref. [9], the road traffic congestion is detected by using a multilevel data fusion method. It also uses a fuzzy logic cluster-based data aggregation scheme. In Ref. [9], Demter-Shafer fusion method is used to determine the congestion on road. Dynamic emergency routing is a major challenge which is addressed in the model that is proposed in Ref. [9]. The authors of paper [10] have proposed an adaptive "neurofuzzy inference system" estimator-based data aggregation scheme for designing of fault-tolerant WSNs. The authors of paper [11] have compared "K-means, Fuzzy C-means, and Fuzzy K-means clustering" based congestion detection techniques. In Ref. [12], the authors proposed an "Adaptive Congestion Aware Routing Protocol" which is able to predict congestion in the network dynamically and determines a safer and reliable route in VANET.

6.3 PROPOSED FUZZY LOGIC-BASED CONGESTION DETECTION AND IoT-BASED EMERGENCY ROUTING

The paper proposes a fuzzy logic-based inference system that detects temporary congestion on-road network. A crowd on road can be detected from the average speed of vehicles on road, sound level, CO_2 level, CO level, temperature, and most important radio frequency level. When there is a crowd on a road, due to mass number of vehicles, the CO_2, CO, and temperature at the particular area increase. Nowadays, every person is using a smartphone, so more the number of persons in a crowd means more is the radio frequency level. All these data can be gathered by different types of sensors, whereas radio frequency detection and spectrum analysis device can detect radio frequency signals transmitted by smartphones in the crowd. The data flow and dynamic routing are shown in Figure 6.1.

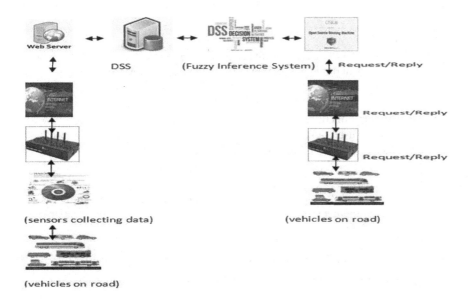

FIGURE 6.1 Data flow and dynamic routing.

A set of sensors, GPS device equipped on a microcontroller unit having Wi-Fi connection, gather on-road traffic data and transmit the data to a traffic monitoring center. The traffic monitoring center is a collective system of web server, database server, decision supporting system, and OSRM server. GPS provides the current geographical location. The web server of traffic monitoring center receives the traffic-related data and current geographical location. All these information are stored in a database server. The decision supporting system works in two phases. In the first phase, a fuzzy logic-based inference system aggregates and analyzes the collected data. In the next phase, it assists the OSRM to find the shortest route for the requested healthcare center.

A microcontroller unit is an integrated unit of various sensors like the LTI's laser sensor, sound sensor, MQ135 sensor, DHT11 sensor, and RFID sensors which collect data like average speed of vehicles, sound, CO_2 and CO, temperature, and radio frequency level. These sensors are integrated on an MCU. The GPS is used to find the current location of emergency vehicles in case of road congestion.

The collected information is then forwarded to a web server through network gateway and Internet. It uses http protocols to do so. The web server then validates the received information and stores valid data in the database server. The decision support system retrieves data in regular interval. It uses fuzzy logic-based inference system to predict for congestion on road. It also forwards the live traffic congestion to the OSRM server. The OSRM server generates the shortest route for the intended location on receiving a request and sends the generated route map to the user.

6.3.1 Fuzzy Logic-based Inference System for Congestion Detection

The decision support system in the first phase uses a fuzzy logic-based inference system for detecting congestion. The fuzzy logic technique works as per the following four steps:

- Fuzzification:
  It is the process of converting a crisp value to fuzzy value. The traffic sensory information that is stored in the database server is crisp input. These crisp inputs are translated to fuzzy inputs by the fuzzification process.

- Membership Functions:
  These are used in fuzzification and defuzzification processes. It maps the crisp value to suitable linguistic variable and vice versa.

- Fuzzy Rule Base:
  They are simple IF-THEN rules that are applied to get fuzzy output for a set of fuzzy input combination.

- Defuzzification:
  The output of the FIS is a fuzzy value. This fuzzy value is defuzzified using a defuzzification technique to obtain a crisp result.

In the proposed model, the input parameters for the FIS are average speed, sound, CO_2, CO, temperature, and radio frequency. The different factors like the slow speed of vehicle, a little noise, more emission of CO_2 and CO, high-temperature difference between city and location of sensors, and high density of radio frequency indicate the presence of crowd or congestion. They are taken as the input parameters, and the congestion level is taken as the output parameter. The linguistic variables for input parameters are low, medium, and high. In the proposed FIS, "triangular membership function" is used for all linguistic variables of input parameters. The linguistic variables for the output parameter (congestion level) are less; moderate, high, and very high. The weighted average method is used for defuzzification to convert the "fuzzy output to a crisp value". The generated fuzzy rules are based on "Sugeno inference system". In Sugeno inference system, if input1 $=x$ and input2 $=y$, then output $z=f(x, y)$. The output z is represented by Equation (6.1).

$$\text{"}z = f(x, y) = ax + by + c\text{"} \tag{6.1}$$

Here, "a, b and c are constants". The output congestion level is set to constant type with constant values $a=b=0$ and $c=0.2$, $c=0.4$, $c=0.7$, and $c=0.8$ for linguistic variables less, moderate, high, and very high, respectively. As $a=b=0$, it is known as zero-order Sugeno model. The various input parameters, their linguistic variables, ranges, and membership functions are presented in Table 6.1.

After detecting the presence of congestion on a road network, the congestion details are updated in the decision support system. The "OSRM" is fed with this updated congestion information. OSRM is a routing engine implemented using C++ to find the shortest path on a road network. It uses multilevel Dijkstra algorithm and OpenStreetMap to compute the shortest path. Here, all the vehicles are assumed to be

TABLE 6.1

Input Parameters for FIS for Emergency Vehicle Routing

Parameters	Type	Linguistic Variable	Range	Membership Function
Average speed	Input	Low	0.0–0.33	Triangular
		Medium	0.27–0.62	
		High	0.58–1.0	
Sound	Input	Low	0.0–0.31	Triangular
		Medium	0.27–0.62	
		High	0.6–1.0	
Carbon dioxide	Input	Low	0.0–0.32	Triangular
		Medium	0.3–0.63	
		High	0.6–1.0	
Carbon monoxide	Input	Low	0.0–0.32	Triangular
		Medium	0.28–0.65	
		High	0.6–1.0	
Temperature difference	Input	Low	0.0–0.31	Triangular
		Medium	0.27–0.64	
		High	0.62–1.0	
Radio frequency	Input	Low	0.0–0.33	Triangular
		Medium	0.3–0.68	
		High	0.63–1.0	

smart vehicles. When any overcrowded road is encountered, congestion is detected automatically with the help of embedded sensors, and the same process is repeated. The plots for the various input and output membership functions are depicted in Figures 6.2–6.8.

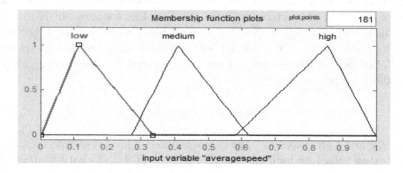

FIGURE 6.2   Average speed membership function.

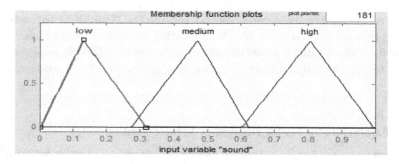

FIGURE 6.3   Sound membership function.

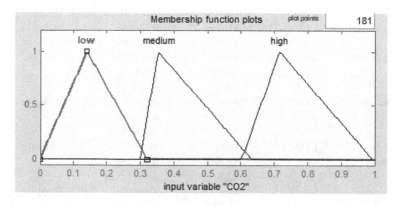

FIGURE 6.4   Carbon dioxide membership function.

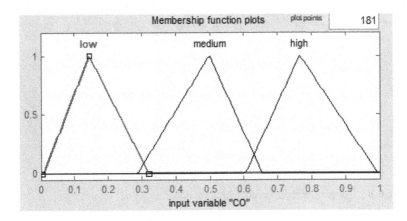

FIGURE 6.5   Carbon monoxide membership function.

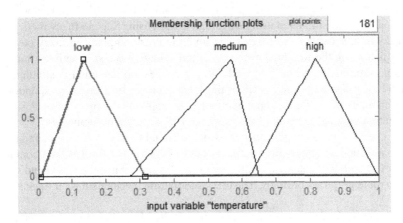

FIGURE 6.6   Temperature membership function.

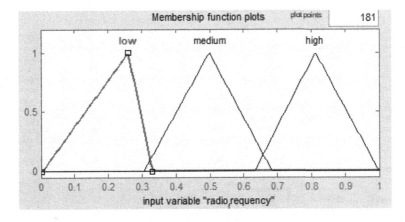

FIGURE 6.7   Radio frequency membership function.

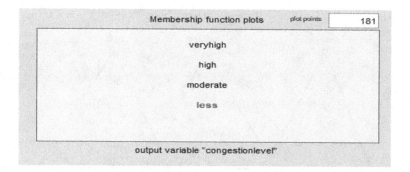

FIGURE 6.8   Congestion level membership function.

6.4   SIMULATION RESULTS

The simulation is done in MATLAB (version-2014a) and Matlab Fuzzy Toolbox. The proposed model is simulated using a FIS that is built using Sugeno fuzzy model with six fuzzy inputs, twenty five fuzzy rules, and one fuzzy output. The fuzzy output is defuzzified using the weighted average method. The FIS is tested with six numbers of input combinations. The proposed FIS for congestion detection is simulated in Matlab Fuzzy Toolbox which provides the value of congestion level corresponding to the given input values for the input parameters by applying the fuzzy rules. A sample input/output simulation result of the proposed FIS is shown in Figure 6.9.

The crisp output is congestion level which is found to be 0.8. In the second phase of decision support system, this output is fed to OSRM. The OSRM then computes the shortest path to the nearest healthcare center from the current location and generates the route graph accordingly. The proposed routing mechanism is outlined below.

a. At the beginning of each simulation round, the FIS estimates the congestion level at the current location.
b. The threshold congestion level is set to 0.65 as simulation parameter.

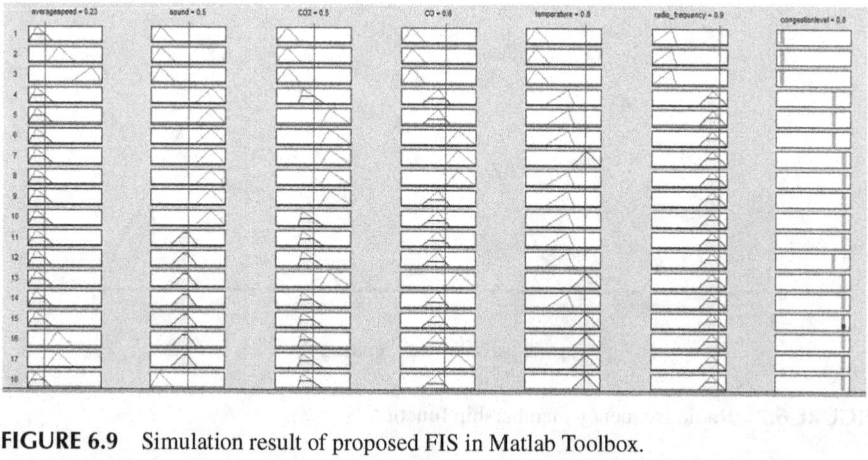

FIGURE 6.9   Simulation result of proposed FIS in Matlab Toolbox.

c. If the FIS output for congestion level is higher than the threshold value, then it is forwarded to OSRM.
d. Else the next round of simulation is conducted.
e. The OSRM gets information about congestion at the location and receives the request for shortest route between the source and the destination.
f. After receiving the request, the OSRM computes and generates the shortest route graph.

6.5 CONCLUSION

In a smart city, routing of emergency vehicles is necessary for safe driving and saving precious lives. The proposed mechanism detects congestion on road due to accidents, construction work, natural disasters, etc. The sensors on a microcontroller unit gather information like average speed of vehicles, sound, carbon dioxide, carbon monoxide, temperature difference between the current location and within the city, and density of radio frequency for a particular location. The Sugeno-based FIS takes these fuzzy inputs and computes the crisp value of the congestion level after defuzzification. Then, the open-source routing algorithm is requested to compute and generate the shortest path to the nearest healthcare center. This technique can save public lives by monitoring traffic and taking decisions smartly which will not only enhance smart healthcare services but also save precious lives.

REFERENCES

1. Naranjo JE, Jimenez F, Serradilla FJ, Zato JG. Floating car data augmentation based on infrastructure sensors and neural networks. *IEEE Trans Intell Transp Syst* 2012; 13(1): 107–14. http://dx.doi.org/10.1109/TITS.2011.2180377.
2. Islam SMR, Kwak D, Kabir MH, Hossain M, Kwak KS. The internet of things for health care: A comprehensive survey. *IEEE Access* 2015; 3: 678–708. http://dx.doi.org/10.1109/ACCESS.2015.2437951.
3. Nadeem T, Dashtinezhad S, Liao C, Iftode L. TrafficView: A scalable traffic monitoring system. In: *Proceedings of IEEE International Conference on Mobile Data Management*, 2004, pp. 13–26. http://dx.doi.org/10.1109/MDM.2004.1263039.
4. Ali K, Al-Yaseen D, Ejaz A, Javed T, Hassanein HS. CrowdITS: Crowdsourcing in intelligent transportation systems. In: *2012 IEEE Wireless Communications and Networking Conference*, 2012, pp. 3307–3311. http://dx.doi.org/10.1109/WCNC.2012.6214379.
5. Handayani AS, Marta Putri H, Soim S, Husni NL, Rusmiasih R, Sitompul CR. Intelligent transportation system for traffic accident monitoring. In: *2019 International Conference on Electrical Engineering and Computer Science*, 2019, pp. 156–161. http://dx.doi.org/10.1109/ICECOS47637.2019.8984525.
6. Tian R, Li S, Yang G. Research on emergency vehicle routing planning based on short-term traffic flow prediction. *Wirel Pers Commun* 2018; 102(2): 1993–2010. http://dx.doi.org/10.1007/s11277-018-5251-2.
7. Jotshi A, Gong Q, Batta R. Dispatching and routing of emergency vehicles in disaster mitigation using data fusion. *Socio-Econ Plan Sci* 2009; 43(1): 1–24. http://dx.doi.org/10.1016/j.seps.2008.02.005.
8. Milojevic M, Rakocevic V. Location aware data aggregation for efficient message dissemination in vehicular ad hoc networks. *IEEE Trans Veh Technol* 2015; 64(12): 5575–5583. http://dx.doi.org/10.1109/TVT.2015.2487830.

9. Zhang L, Gao D, Zhao W, Chao HC. A multilevel information fusion approach for road congestion detection in VANETs. *Math Comput Modelling* 2013; 58(5): 1206–1221. http://dx.doi.org/10.1016/j.mcm.2013.02.004.

10. Acharya S, Tripathy CR. An ANFIS estimator based data aggregation scheme for fault tolerant Wireless Sensor Networks. *J King Saud Univ – Comput Inf Sci* 2018; 30(3): 334–348.

11. Mohanty A, Mahapatra S, Bhanja U. Traffic congestion detection in a city using clustering techniques in VANETs. *Indones J Electr Eng Comput Sci* 2019; 13(3): 884–891. http://dx.doi.org/10.11591/ijeecs.v13.i3.pp884-891.

12. Giripunje LM, Vidyarthi A, Shandilya SK. Adaptive Congestion Prediction in Vehicular Ad-hoc Networks (VANET) using Type-2 Fuzzy Model to Establish Reliable Routes. https://doi.org/10.21203/rs.3.rs-458059/v1.

7 A Fog Cluster-Based Framework for Personalized Healthcare Monitoring

Jeyashree G. and Padmavathi S.
Thiagarajar College of Engineering

CONTENTS

7.1 INTRODUCTION

Cloud computing has taken up the research area and the IT market with its inherent features of scalability, interoperability, computing on demand, and availability. Pay-as-you-use policy has been made available to almost all researchers and professionals at an affordable cost [1]. Significant services of the cloud that made the success possible are the Infrastructure as a service, Platform as a service, and Application as a service which in today's scenario has to be extended to Everything as a service.

The application developer or researchers need not worry about the Infrastructure or the platform need. Once the data is ready, the need for computing and communication facilities is requested which can be taken up as a service. There have been several works stating the use of the cloud for IoT applications [2–4]. With the list of advancements and advantages, there are some major pitfalls in the case of IoT applications relying on the cloud platform. A large amount of data have to be analyzed in the near real time, and the decision has to be sent back to the IoT device for actuation. Sending a large amount of data to the cloud at a remote location leads to very large consumption of network bandwidth [1]. In addition to the bandwidth consumption, sending and receiving data and decisions from a remote server induce communication latency and lead to the overall latency of the hosted application.

The latency in decision-making and communication may lead to working on the expired data from the sensors, thereby leading to inaccurate decisions and a reduction in the application's performance. Thus, to overcome these limitations of cloud processing, selective computations can be offloaded to the edge of the network and minimal computations can be done on the end devices themselves. Thus, the latency-aware computations can be performed at the edge and the computationally intensive tasks can be pushed to the cloud and permanent storage can be done at the cloud for further intelligence on the application and collection of data in the respective area. The use of the intermediary fog layer can further help filter the data sent to the cloud, which ensures the optimal use of bandwidth.

Through the years, the world has moved on to the fantasy of automation in almost every activity of our life. The fantasy has almost become a reality with the introduction of Artificial Intelligence and the availability of a huge set of real-time data to analyze with. The data collection and availability were once a whimsy which have now become very much possible with the help of sensors and IoT devices and the availability of communication and computation modules at an affordable rate. One of the major characteristics of this data is that its lifetime is very much limited and is of no use after expiry [5]. For example, the data from the temperature sensors of an environment has to be analyzed immediately with any latency to bring out a meaningful actuation. The use of IoT devices to assess a large environment requires a large number of sensors depending on the area of analysis. As the number of devices increases, the volume and velocity of data also increase, which bring out a major challenge of the real-time analysis of these data within a short span of time. These data have to be sent for analysis, and the Infrastructure being used at large is the cloud platform.

Chronic obstructive pulmonary disease (COPD) refers to a group of diseases that causes blockage in airflow and causes breathing-related problems [6]. Statistics says that COPD is the third leading cause of death in the United States, and almost 15.7 billion US people were diagnosed with COPD by 2014. Medical experts say that there is no cure for COPD, but it can be treated. Although smoking and tobacco are the major risk factors for COPD, more women are affected by COPD than men and it is the highest in people aged above 65 years. Hence, monitoring the elderly with COPD can avoid intensifying their disease and can extend their lifetime. This paper explains the benefits of involving fog computing for pre-processing health data in monitoring patients with COPD and thereby propose an idea of connecting IoT devices to the fog layer rather than to the cloud which in turn reduces the latency, bandwidth, and energy consumption in a significant manner.

The main contributions of this paper are as follows:

- Presenting the importance of fog computing in case of handling latency-sensitive IoT applications in the healthcare sector

- Proposing fog clusters with multiple nodes in the fog layer to carry out the analytics and thereby provide a faster response
- Evaluating the proposed Infrastructure with a suitable application scenario in healthcare
- Simulating the whole system of our healthcare use case and providing results comparing the cloud and fog in terms of latency and energy consumption

The main objective of this paper is to impose the benefits of fog computing in healthcare IoT, and the rest of the paper is structured as follows: Section 7.2 elaborates preliminaries of the fog computing paradigm. Also, the healthcare sector applications and how they are being benefited from fog computing are described in this section. In Section 7.3, our proposed Infrastructure is demonstrated with a suitable healthcare use case. The results and discussions are presented in Section 7.4, and Section 7.5 concludes the paper.

7.2 PRELIMINARIES

This section elaborates the detailed study of fog computing, its significance, and how it has been utilized in many healthcare applications. The adoption of fog computing in significant applications like healthcare has numerous advantages leading to promising solutions in many healthcare cases. This makes the system trustworthy and hence the acceptance of the system by the stakeholders keeps increasing.

7.2.1 FOG COMPUTING

The rapidly increasing number of sensors that tries to connect to cloud infrastructure may not be suitable for latency-aware applications like healthcare. Also, in many cases, storing medical data in a third-party server is not recommended due to privacy issues. In the case of network failure and emergencies, data availability may be at risk. The fog layer, which forms an intermediary layer between the cloud and devices, can rescue such problems. Fog can be seen as an extension of the cloud but much near to the devices on one's own premises. This layer doesn't take up heavy computing, and the resources are constrained too. But the processing taken up by the fog layer can reduce latency at large and also give high-level privacy to the data being processed in the application. The cloud can act as a supporting layer to perform heavy computations and for permanent data storage.

A common application of fog is that it allows the doctor to make immediate choices with the data available and based on the analytics in the sensor data. The data is also confidential and decreases the latency and bandwidth when compared to a cloud-only infrastructure. The fog layer can be visualized as shown in Figure 7.1.

7.2.2 FOG COMPUTING FOR HEALTHCARE

Some of the vital applications of fog in healthcare are given in Figure 7.2.

There have been several improvements in the area of healthcare since the era of sensors. The sensors collect data 24/7 from the patients, and the focus on monitoring

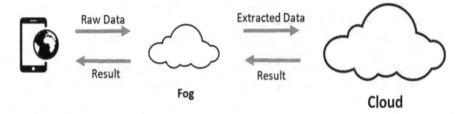

FIGURE 7.1   Basic fog architecture.

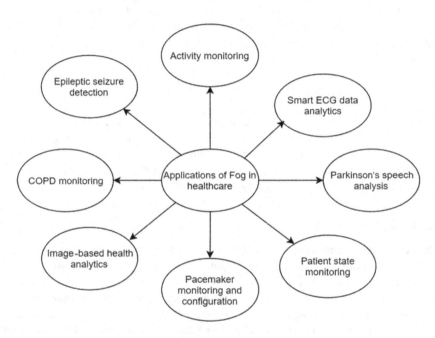

FIGURE 7.2   Vital applications of fog in healthcare.

should not be neglected. Thus, an automated system to analyze the data has always been in research. The model developed by [7] introduces the importance of monitoring without physical intervention. As digitalization improved, there was an era of body area networks. The uses of ECG, temperature, heart rate, and other vital parameter sensors were continuously monitored with intelligence embedded in the sensors too [8]. Some actuators can be alarms or medicine injectors that work based on the sensor data.

The actuators and sensors constantly communicate with the vital parameters of patients and are immediately notified to the authority also [5]. There are sensors that are integrated and perform multiple functionalities too. But in case of places where the network connectivity is low or limited like military bases or secluded areas, the data has to be analyzed remotely and the report has to be produced in the least

TABLE 7.1
Literature with Fog Layer for Healthcare

Literature	Aim	Components Used	Outcomes
[9]	Home monitoring or hospitalization	NodeMCU, basic sensors	Cost-effective and edge inference
[10]	Health assistance	Cloud server, wearable sensors, robot	Highly efficient but least affordable. Abstractions provided clearly in the robot and cloud processing
[11]	Deep learning (DL)–based cardiac disease analytics	Sensors, fog gateway, routers, cloud server	Efficient DL-based inferences. But not suitable for embedded system implementations
[12]	Daily life monitoring	Sensor network, cloud server	Fog-based daily monitoring system for smart health assessments

possible time [13]. The hand-over in the network and the remote location of the server can inhibit the performance of the application, and the introduction of an intermediary fog layer can help in performance improvement, especially latency [14]. The best feasible solution to handle the sensor data can be the fog infrastructure that makes the application highly reactive. The fog does not replace the cloud but acts as an assistive layer to improve the application's performance.

The fog layer can handle lightweight machine learning models and give intelligent actuations based on the input streaming data in real time [15]. The fog can also be visualized as a peer-to-peer model between the device and local server. Since the fog nodes are close to the things/IoT devices, the performance is elevated, especially in terms of latency and bandwidth utility. Some of the notable literatures are listed in Table 7.1.

7.3 PROPOSED ARCHITECTURE

Though fog computing has several benefits of its own, it has a few challenges too as described in Ref. [16] such as its structural issues and service-oriented and security aspects. The challenges of structural and service-oriented issues are often based on the constraints in fog nodes. Fog nodes are not like conventional cloud data centers and the development of large-scale applications in such resource-constrained nodes is difficult. The cluster-based fog approach proposed in this paper can be one of the solutions for solving the problems mentioned earlier. With the proposed fog cluster, we get the possibility of grouping multiple resource-constrained fog nodes, and we end up having a relatively better resource-enriched node to carry out more tasks in parallel. Deploying workload on the fog nodes is another challenge as described in Ref. [17].

The deployment of an application into a fog cluster reduces the time taken to individually deploy its modules among multiple fog nodes. By deploying an application to a fog cluster, through intercluster communication, application modules are distributed across fog nodes in the cluster and carry out processing in parallel.

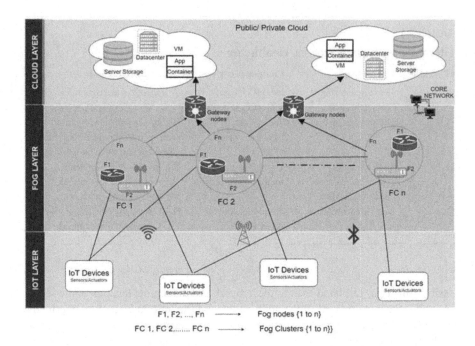

FIGURE 7.3   Fog cluster architecture.

Figure 7.3 shows the proposed Infrastructure with three layers namely: IoT layer, fog layer and cloud layer. The connection between IoT and fog layer requires only low-level connectivity such as Zig-Bee and Bluetooth, whereas cloud and fog require LAN or Wi-Fi. Clustering of fog nodes is the proposed technique in this paper. It comes up to be a wonderful solution to meet the stringent latency requirements of an application and ensure high availability. Logically, a fog cluster is a group of fog nodes intended to divide the tasks and process them in parallel collaboratively. A masterless architecture-based distributed fog cluster is created. It has no central coordinator; hence, nodes can be easily added and removed from the cluster. The time series IoT data is gathered and distributed across the cluster of nodes. Every node in the cluster has a minimal capacity to perform pre-processing of data such as data gathering, filtering, aggregation, etc. On failure of a fog node, the task is automatically assigned to another fog node in the cluster, thus avoiding a single point of failure. Even when all the fog cluster nodes fail, the module is assigned to the next cluster since intercluster and intracluster communication is possible in the fog layer.

7.3.1   CLUSTER-BASED ARCHITECTURE

Our proposed architecture is a cluster-based architecture where the cluster of devices acts as the fog layer. There can be situations where the network and bandwidth availability are limited. Thus, processing the large sensor data that are streaming can be difficult. We have created a cluster of devices that takes up the data and divides the tasks to be completed. Thus, there is an efficient use of resources and also multiple tasks are handled with ease. The optimal use of the fog layer can be in situations as follows:

- When there are hardware limitations in the number of cores and memory
- Limited connectivity and latency
- Restrictions on maximum connections or instances
- Cost to model and node placements

As given in Figure 7.4, a cluster of nodes parallelizes the task that the cluster head receives. Initially, the open-ssh is installed to avoid frequent authentications to execute the tasks. This creates the ssh keys among the nodes in the cluster. On creating the cluster, the IP address of the cluster nodes is added to the /etc/host file in the cluster head. This file helps establish a secure connection among the nodes. The following commands are desirable for the IPv6-capable hosts ::1 ip6-localhost ip6-loopback fe00::0 ip6-localnet, ff00::0 ip6-mcastprefix, ff02::1 ip6-allnodes, ff02::2 ip6-allrouters. The configurations of the cluster created are illustrated in Figure 7.5. It also explains the steps involved in creating the cluster.

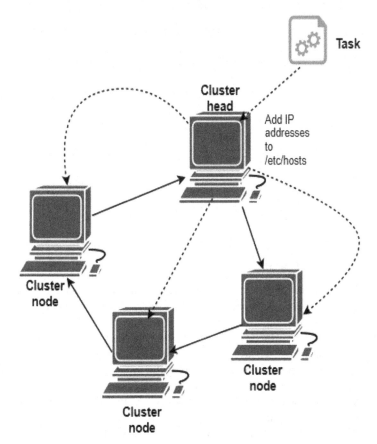

FIGURE 7.4 Cluster creation.

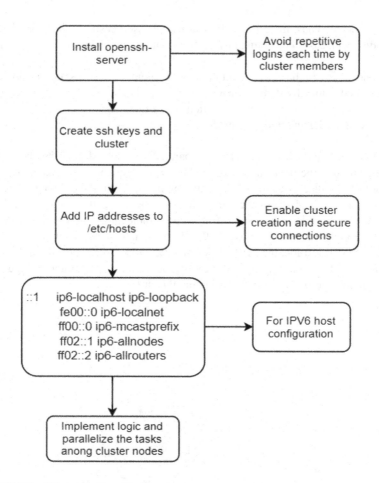

FIGURE 7.5   Cluster configuration.

7.3.2   EXPERIMENTAL SETUP

To evaluate the proposed Infrastructure, we simulated a fog environment using iFog-Sim [18]. It is built on top of the standard CloudSim toolkit that has been widely used for simulating various computing paradigms. The application scenario mentioned in Section 7.4. (a) is implemented in the proposed Infrastructure. In the simulated environment, we considered three layers, such as the IoT layer with sensors and actuators. The fog layer with 'n' number of fog clusters with multiple fog nodes, four fog nodes are simulated in one fog cluster. The assumption is that the cluster is highly available, fault-tolerant, and free from Byzantine attacks—finally, the cloud layer with data centers and virtual machines.

Figure 7.6 shows the fog cluster created in the fog layer based on a Ring topology, and the nodes in the fog cluster are homogeneous. We employ a hashing technique to distribute tasks among the fog nodes in the cluster uniformly. When time series data

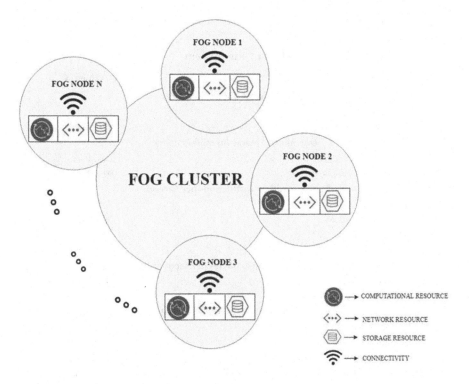

FIGURE 7.6   Fog cluster.

is fed as input from the IoT layer, it is first assigned to a cluster and any node takes responsibility and distributes the tasks of filtering, aggregation, and analyzing across other nodes in the cluster. Failure of a node is identified with a timestamp value. After the timestamp, if there is no response from that particular node, the task of that node is redirected to the next node in the cluster. Here, the timestamp is set to 0.2 ms.

The simulation parameters for the overall Infrastructure are listed in Table 7.2.

7.3.3 APPLICATION SCENARIO

When it comes to healthcare, fog computing is the most valuable for patient monitoring and healthcare applications in rural areas where access to a hospital is not so easy. Hence, this paper considers an application scenario of monitoring a patient suffering from COPD. Pulse rate (PR) and oxygen saturation (SpO2) in this COPD patient monitoring system are continuously monitored. PR can be monitored through PR sensor and SpO2 is monitored through pulse oximetry. When there is a sudden rise or fall beyond a threshold value in either PR or SpO2, the patient may find it hard to breathe and it is considered an emergency. A fall of the SpO2 range below the normal may cause hypoxemia, a respiratory disorder. Note that the threshold value differs from patient to patient depending on their health. This application scenario is latency-sensitive as it deals with the life of a patient, and any slower response may badly influence the patient's health condition. So, basically, our application scenario is the patient's vital sign analysis and detection of drastic fluctuations (emergency).

TABLE 7.2

Configuration Parameters of the System

Parameters	Values
Uplink Latency: (in milliseconds)	
IoT layer to Fog layer	20–30 ms
Fog layer to Cloud layer	80–100 ms
Downlink Latency: (in milliseconds)	
IoT layer to Fog layer	25–30 ms
Fog layer to Cloud layer	80–100 ms
CPU: (in Million Instructions per Second)	
Sensor module	350 MIPS
Coordinator module	300 MIPS
Analyzer module	200 MIPS
Actuator module	250 MIPS
Bandwidth: (in kilobytes per second)	
Gateway to cloud link	10,000 kbps
Fog node to the gateway	10,000 kbps
Processing Time: (in milliseconds)	
Sensor module	20–40 ms
Coordinator module	15–20 ms
Analyzer module	150–200 ms
Actuator module	50–80 ms

Diagnosing such emergency conditions at the right time could save their lives. Analysis of a drastic change in attribute above or below the threshold range is carried out in the fog node. Faster processing of this real-time data and immediate response during an emergency by fog computing would ensure the true potential of having this done in the fog layer rather than sending them to the cloud. Such real-time processing requires that the application be hosted close to the user, making the application a typical use case for fog computing (Figure 7.7).

The COPD patient monitoring application has four modules: the sensor, coordinator, analyzer, and actuator. The sensor modules collect the PR and SpO2 data continuously from the patient through wearables. The collected raw data are sent to the coordinator module. The coordinator module filters consistent data and forwards them to the analyzer module. In the analyzer module, the actual analysis and detection of vital signs occur by continuously comparing the data with the threshold range. In any encounter of fluctuations below or above the threshold range, the analyzer module sends an alert to the actuator module. The analyzer module is also responsible for storing the data back in the cloud depending on the need and permanence of the data. The actuator module alerts the caretaker as well as sends the vital sign values to the physician of that patient. It is considered that each node in the fog cluster takes responsibility for each module.

Algorithm 7.1 shows the COPD patient monitoring system that inputs time series data from the sensors (wearables). The input dataset is synthesized with multiple sensor readings such as pressure, PR, SpO2, and temperature. Since COPD patients have deviations in their PR and SpO2 levels, it is taken to consider for threshold. The threshold value for each sensor differs from patient to patient depending on age, body size, medication use, and any disease. Here, we have considered a normalized threshold value range for adults between 60 and 100 beats per minute (bpm). The normal range of SpO2 is 95%–100%; however, for COPD patients, these ranges may not apply. Hence, we normalize the SpO2 threshold range for COPD patients as 90%–95%. When a data is found below or above the threshold range, the analyzer module sends a notification to the actuator module. The actuator module notifies the caretakers as well as physicians.

Algorithm 7.1. COPD Patient Monitoring

Input: Measurement of PR and SpO2 of patient through sensors
Output: Alert notification on identifying criticality
Steps:
  1. Set up the Fog Environment with three-level computing hierarchy
  2. Initialize fog nodes with specified attributes
  3. Establish connection between all the components
  4. Input the sensor data from the wearables
  5. Distribute the following tasks across the fog nodes
  6. Set up threshold value as:
       a. PR for the pulse rate sensor
       b. SpO2 for the oxygen saturation sensor
  7. If {PR >60 && PR <100} && {SpO2 > 95 && SpO2 <100},
     then user is in normal stage
              Else If {PR < 60 || PR > 100} || {SpO2 < 95 && SpO2 >100},
              then user is in critical
                       a) Notification is sent to the **physician as well caretaker**
                          via the in-app notification to the application
              Else data are stored in the cloud for future analysis

  8. Exit

7.4   RESULTS AND DISCUSSIONS

All the environments mentioned in the previous sections were simulated in the fog environment. The physical entities such as the fog devices, sensors, and actuators need to be created and their capacities and configurations should be specified. We have set the configurations as mentioned in Table 7.2. The connecting links are also specified in detail about which nodes are connected to which other nodes. The data will be transmitted from one to another, and the transmission rates have to be set before the simulation. The number of resources (CPU and RAM) required to process the nodes must be specified.

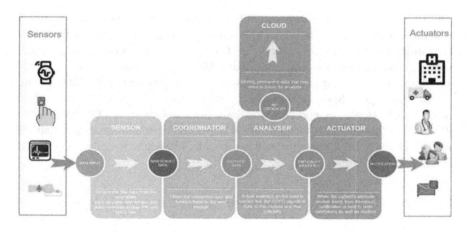

FIGURE 7.7   COPD patient monitoring system with its modules.

The application modules are then modeled and further placed on the nodes in the simulator. The latency-sensitive application, COPD in our case study, involves human–machine interaction. To monitor COPD patients continuously, the patient needs to wear a PR sensor for collecting PR and a pulse oximeter for collecting SpO2. This application performs real-time processing of patient data and calculates the criticality using a threshold for each value.

The results of this simulation show how placing the application modules in the fog
cluster impacts the latency and energy consumption rather than placing them on discarded fog nodes. For testing the simulation with cloud-only and fog placement, we have varied the number of gateways by keeping the number of sensors and actuators constant. For cloud placement of applications, we have kept only two levels of hierarchy such as the sensor layer and cloud layer connected through a gateway and for fog placement, and further, the hierarchy is raised to three levels, such as sensor, fog, and cloud layer, each connected from one to another through a gateway.

The performance of an application is dependent on the time it takes to connect from one node to another. This simulation shows latencies for connecting from one level of nodes/devices to the other level in Table 7.2.

The comparison of delay between cloud-only and fog placement is shown in Figure 7.8. This is because in cloud placement it needs to switch its processing between distant nodes in the cloud, whereas in fog, the process is carried out at nearby nodes. With the fog cluster, the delay can be further reduced since all the application modules in an application can be processed with faster responses by placing them in a single cluster.

Similarly, Figure 7.9 shows the comparison of energy consumed during different placements of applications. While fog nodes are resource-constrained, some tasks may require resource-intensive nodes. By offloading tasks among the group of multiple fog nodes, the computational complexity can be reduced. The amount of energy spent to connect or communicate from one node to another is reduced, hence reducing the computational capacity.

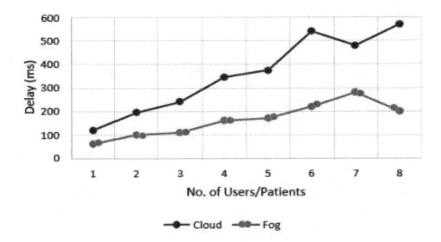

FIGURE 7.8 Performance of fog with respect to latency.

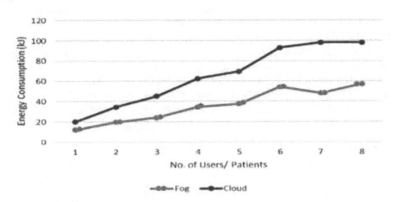

FIGURE 7.9 Performance of fog with respect to energy consumption.

7.5 CONCLUSION

In this paper, the concept of fog computing for healthcare IoT was discussed, and the deployment of the application on a fog cluster rather than fog nodes was presented. The proposed system was also simulated and evaluated with a COPD patient monitoring system, which is a latency-sensitive application. An algorithm to analyze the criticality of the patient's state based on a threshold value was also presented. And the simulation results show that latency is reduced to a greater extent than that of executing them on the cloud infrastructure. The results also show that energy consumption is reduced as the process is handled at a closer distance. However, in the future, we plan to implement this system for various other real-time applications like smart cities, smart agriculture, etc. The fog cluster can be implemented using multiple Raspberry Pi as fog nodes, and our use case can be implemented. Further intelligence can be incorporated to enhance decision-making.

ACKNOWLEDGMENT

The author would like to express her gratitude to her host institution, Thiagarajar College of Engineering, for funding this research through the TCE Research Fellowship (TRF) program Ref. No. TCE/SAVITHA/TRF/2020/2600/01.

REFERENCES

1. P. O'Donovan, C. Gallagher, K. Leahy, D. T. O'Sullivan, A comparison of fog and cloud computing cyber-physical interfaces for industry 4.0 real-time embedded machine learning engineering applications. *Computers in Industry* 2019, 110, 12–35..
2. G. Fortino, A. Guerrieri, W. Russo, C. Savaglio, Integration of agent-based and cloud computing for the smart objects-oriented iot, in: *Proceedings of the 2014 IEEE 18th International Conference on Computer Supported Cooperative Work in Design (CSCWD)*, IEEE, Hsinchu, 2014, pp. 493–498.
3. C. Huo, T.-C. Chien, P.H. Chou, Middleware for iot-cloud integration across application domains. *IEEE Design & Test* 2014, 31(3), 21–31.
4. A. Botta, W. De Donato, V. Persico, A. Pescapé, On the integration of cloud computing and internet of things, in: *2014 International Conference on Future Internet of Things and Cloud*, IEEE, Barcelona, 2014, pp. 23–30.
5. M.H. Riaz, U. Rashid, M. Ali, L. Li, Internet of things based wireless patient body area monitoring network, in: *2017 IEEE International Conference on Internet of Things (iThings) and IEEE Green Computing and Communications (GreenCom) and IEEE Cyber, Physical and Social Computing (CPSCom) and IEEE Smart Data (SmartData), Exeter*, 2017, pp. 970–973, doi: 10.1109/iThingsGreenCom-CPSCom SmartData.2017.180.
6. Centers for Disease Control and Prevention, CDC 24/7: Saving lives, Protecting people, National Center for Chronic Disease Prevention and Health Promotion, Division of Population Health.
7. V. Gupta, A. Ingle, D. Gaikwad, M. Vibhute, Patient health monitoring and diagnosis using IoT and machine learning, in: Ranganathan G., Chen J., Rocha Á. (eds) *Inventive Communication and Computational Technologies. Lecture Notes in Networks and Systems*, vol 145. Springer, Singapore, 2021. doi: 10.1007/978-981-15-7345-3_24.
8. M. R. Ruman, A. Barua, W. Rahman, K. R. Jahan, M. Jamil Roni, M. F. Rahman, IoT based emergency health monitoring system, in: *2020 International Conference on Industry 4.0 Technology (I4Tech)*, 2020, pp. 159–162, doi: 10.1109/I4Tech48345.2020.9102647.
9. H. Ben Hassen, N. Ayari, B. Hamdi, A home hospitalization system based on the Internet of things, Fog computing and cloud computing. *Informatics in Medicine Unlocked* 2020, 20, 100368.
10. M. Pham, Y. Mengistu, H. Do, W. Sheng, Delivering home healthcare through a Cloud-based Smart Home Environment (CoSHE). *Future Generation Computer Systems* 2018, 81, 129–140.
11. S. Tuli, N. Basumatary, S.S. Gill, M. Kahani, R.C. Arya, G.S. Wander, R. Buyya, HealthFog: An ensemble deep learning based Smart Healthcare System for Automatic Diagnosis of Heart Diseases in integrated IoT and fog computing environments. *Future Generation Computer Systems* 2020, 104, 187–200.
12. O. Debauche, S. Mahmoudi, P. Manneback, A. Assila, Fog IoT for health: A new architecture for patients and elderly monitoring. *Procedia Computer Science* 2019, 160, 289–297.
13. N. Patii, B. Iyer, Health monitoring and tracking system for soldiers using Internet of Things(IoT), in: *2017 International Conference on Computing, Communication and Automation (ICCCA)*, Greater Noida, 2017, pp. 1347–1352, doi:10.1109/CCAA.2017.8230007.

14. S. P. Ahuja, N. Deval, From cloud computing to Fog computing. *International Journal of Fog Computing* 2018, 1(1), 1–14. doi:10.4018/ijfc.2018010101.
15. Q., Xie, G. Wang, Z. Peng, Y. Lian, Machine learning methods for real-time blood pressure measurement based on photoplethysmography, in: *2018 IEEE 23rd International Conference on Digital Signal Processing (DSP)*, 2018, pp. 1–5. doi:10.1109/ICDSP.2018.8631690.
16. R. Mahmud, R. Kotagiri, R. Buyya, Fog computing: A taxonomy, survey and future directions, in: Di Martino, B., Li, K-C., Yang, L. T., Esposito, A. (eds) *Internet of Everything*, pp. 103–130. Springer, Singapore, 2018.
17. N. Wang, B. Varghese, M. Matthaiou, D.S. Nikolopoulos, Enorm: A framework for edge node resource management. *IEEE Transactions on Services Computing* 2017, 13(6), 1086–1099.
18. H. Gupta, A. Vahid Dastjerdi, S. K. Ghosh, R. Buyya, iFogSim: A toolkit for modeling and simulation of resource management techniques in the Internet of Things, Edge and Fog computing environments. *Software: Practice andExperience* 2017, 47(9), 1275–1296.

8 QTR
QoS-Aware Traffic Rerouting in OpenFlow-Based Software-Defined Networks

S. K. Nivetha
Kongu Engineering College

N. Senthilkumaran
Vellalar College for Women

M. Geetha and R. C. Suganthe
Kongu Engineering College

CONTENTS

8.1   INTRODUCTION

Software-Defined Networking (SDN) is an intelligent networking [1] concept to manage networking nodes remotely through programming with a split architecture decoupling control and data planes. Studies reveal that, in a large-scale network, the frequency of failure is more on the forwarding plane because a link can fail every 30 minutes [2] on average. As failure on the forwarding plane leads to data loss which is not preferable in reality, failure detection and recovery have become mandatory in such networks. Moreover, today's internet applications have their own demand of Quality-of-Service (QoS) requirements.

QoS [3] is defined as the ability of the network to provide the service requirements mandated by the applications while traffic routing. This is especially important for multimedia-related real-time applications like video conferencing, Voice-over IP (VoIP), distance learning and online interactive gaming. Different applications have different requirements (i.e.) video conferencing mandates a certain bandwidth whereas VoIP mandates minimum/no delay and no jitter. Hence, it is necessary for the underlying network to segregate the traffic flows accordingly and provide the QoS mandated by them. With the traditional best-effort Internet architecture, it is still impossible and an ongoing area of research to provide the aforementioned services. But with the programmable and flexible SDN architecture, it is possible to manage and allocate the network resources dynamically.

In recent years, SDN, as an emerging technology, is providing solutions to large-scale cloud data centres and big data business sectors. Network functions of those large-scale cloud systems are distributed across various locations from the central server [4]. Moreover, the controller functionality might also have been distributed and the real-time traffic is delay-intolerant and we could not apply controller-initiated reactive restoration-based recovery as it leads to more delay. Proactive protection-based failure recovery approaches are proven to result in less delay as they use preconfigured backup paths to switch to the alternate path without controller intervention in case of failure. Hence, this work proposes a proactive failure detection and recovery approach with QoS to reroute the traffic quickly and automatically in OpenFlow-based SDN.

Moreover, in order to compute the backup paths, the controller must be enabled with an effective multipath algorithm. The braided multipath scheme is a viable alternative to the node-disjoint multipath scheme in which the alternate paths computed are only partially node-disjoint, thus providing relaxation on 100% node-disjointedness. Furthermore, the alternate paths computed using a braided scheme are not deviating much from the primary as they are formed by changing one or few nodes in the primary path. Because of the effectiveness of the proposed braided multipath scheme, this work uses it to compute the alternate backup paths in the controller.

For failure detection, this work aims to use the successful detection technique called Bidirectional Forwarding Detection (BFD) [5] technique. It is useful for network administrators to detect forwarding path failures at a uniform rate. Hence, it can be recovered and reconstructed in a consistent and predictable manner. BFD can be applied for any type of link between systems, to detect failures. Since it is proven as a simple and fast method for detecting a failure in the recent literature, this work aims to apply this technique for effective failure detection. Consequently, QoS-aware Traffic Rerouting (QTR) is formulated with BFD as a failure detection protocol and proactive protection approach for failure recovery that uses braided multipath to configure backup paths. The working and backup paths that satisfy quality attributes/ metrics mandated by the real-time applications are chosen by the controller.

8.2   RELATED WORKS

Civanlar et al. proposed an architecture for supporting QoS flows in an OpenFlow-based SDN environment. This work [6] considers delay and packet loss. The authors used linear programming-based path calculation for video traffic and the shortest path algorithm for the best-effort traffic. An SDN-based QoS control framework [7] has been proposed by Kim et al. which allows network devices to program QoS parameters. But, the extensions implemented on SDN were not given by the authors. Also, the evaluation of the performance of the proposed framework was not done by the authors.

Egilmez et al. have presented an effective controller design [8] that supports end-to-end QoS, particularly for real-time multimedia delivery. As an extension, Egilmez et al. have proposed a framework [9] for adaptive video streaming in SDN. This work provides an optimization framework for the control layer in order to enable dynamic QoS in an OpenFlow-based SDN. The authors proposed different mechanisms for streaming video data and evaluated with respect to performance and cost. Particularly, the framework supports scalable encode videos with two different QoS levels. Egilmez and Tekalp have proposed distributed QoS architectures [10] for multimedia streaming in an OpenFlow-based SDN.

Wendong et al. have proposed an Autonomic QoS management mechanism [11] for Software-Defined Networks (AQSDN). The detailed architecture of AQSDN is presented. The authors stated that various QoS features can be configured simply by extending the existing OpenFlow and OF-Config protocols. They have built a Packet Context-aware QoS (PCaQoS) model in order to improve the network QoS. The PCaQoS model is implemented and compared with the DiffServ model. Based on several analyses, the authors concluded that this work shows better results for simple network environments and it requires further exploration of larger networks which is the actual demand today.

Govindarajan et al. proposed a QoS Controller [12], the Q-Ctrl, with five major operations including queue creation, QoS flow addition, QoS flow modification, QoS flow deletion and queue deletion. It has been proposed particularly for cloud infrastructure.

Jinyao et al. proposed Hi-QoS [13] which uses different queues for different kinds of applications. It also addresses failure recovery using a multipath scheme. The components proposed by this system are the differentiated service and multipath routing component. Differentiated service component is used to distinguish different application services based on the IP address of the source by the controller and provide bandwidth guarantees accordingly. Multipath routing component is used to compute multiple paths that meet defined quality requirements in order to cater failure recovery. Upon failure, the optimal path among the multiple paths identified is chosen based on the current network statistics. However, this work has some overhead as it involves posing queuing mechanisms for distinguished services. Xu et al. proposed a bandwidth- and energy-aware flow scheduling technique [14] in SDN. Zhang et al. proposed a latency monitoring method [15] in SDN. But, they did not focus on failure recovery.

Karakus and Durresi have surveyed several QoS-related works [3] on an SDN paradigm and categorized them into seven major streams. They are Multimedia flow-based routing mechanisms, inter-domain routing mechanisms, resource reservation mechanisms, queue management and scheduling mechanisms, quality-of-experience-aware mechanisms and other QoS centric virtualization-based approaches. The authors listed and briefed several existing solutions under each category for the benefit of the researchers in this domain.

The authors also tabulated various existing works based on the QoS model it supports like hard-QoS and soft-QoS. And further, they are categorized based on the quality metrics that are considered in that work. Also, for each work, the simulation environment used and the plane in which they are implemented are also tabulated. This work will serve as a guide for the researchers of this domain. Bagaa et al. proposed a QoS-aware routing scheme [16] in SDN using multipath. In that, the authors used three solutions namely full-path recomputation based on linear programming, heuristic-path recomputation based on linear programming and partial-path recomputation based on Dijkstra's algorithm. But, this work does not focus on failure recovery.

Senthilkumaran and Thangarajan [17] proposed a failure recovery method by automatic traffic rerouting (ATR). In this work, failure recovery is automatic using backup paths. But, these backup paths are computed based on the shortest path only. The authors extended their work and proposed memory- and load-aware traffic rerouting (MLTR) [18]. In this, these backup paths are calculated in order to reduce the Ternary Content-Addressable Memory (TCAM) occupancy by the flow rules, and additionally, the flow rules are chosen so as to reduce the load on the switches.

The studies above are evident that forwarding plane failures are having a notable impact on SDN, and it necessitates an effective failure recovery mechanism. A proactive approach rather than reactive is preferred in failure recovery to considerably reduce the recovery time. Additionally, any recovery mechanism may also consider QoS-related issues, which is a mandate for real-time communication. Hence, ATR is extended to provision QoS while choosing the primary and backup paths.

8.3 METHODOLOGY

8.3.1 QoS Metrics

Deciding the QoS metrics according to the flow requirement becomes an intricate task [19]. Generally, any QoS metric can be categorized [20] into additive, concave and/or multiplicative.

Additive: the metric's collective value over the path. E.g. delay, jitter, cost, hop count/distance.

Concave: the metric's minimum value over the path. E.g. bandwidth.

Multiplicative: the metric's product of values over the path. E.g. probability of packet loss.

Other metrics like throughput, packet loss ratio (PLR) and packet delivery ratio are categorized into separate classes as they are calculated based on the success ratio.

Let Qm(v_1, v_2) be a QoS metric for link(v_1, v_2). v_1, v_2, v_3, ..., v_i, v_j represent network nodes/vertices. Let P be the path in the network, $P = (v_1, v_2, v_3, ..., v_i, v_j)$. Let Qm($P$) be the QoS metric of the path.

If the metric is additive, then

$$Qm(P) = Qm(v_1, v_2) + Qm(v_2, v_3) + \cdots + Qm(v_i, v_j)$$

If the metric is concave, then

$$Qm(P) = \min\{Qm(v_1, v_2), Qm(v_2, v_3), ..., Qm(v_i, v_j)\}$$

If the metric is multiplicative, then

$$Qm(P) = Qm(v_1, v_2) * Qm(v_2, v_3) * \cdots * Qm(v_i, v_j)$$

The general requirements of any real-time application would be a minimum delay and maximum available bandwidth. This work considers end-to-end delay and effective bandwidth of the path while choosing working and backup paths. They are defined as follows:

8.3.2   END-TO-END DELAY

As end-to-end delay is calculated as the cumulative value of all link delays of the path, it is an additive metric. It is generally calculated as the time taken to successfully deliver the data from source to destination and it is a network-related metric [21]. End-to-end delay is a compounded value as it is inclusive of propagation, queuing, processing and transmission delay encountered at each node and/or link of the path. Suppose if there are 'n' nodes and '$n-1$' links in the path, the end-to-end delay is calculated as given in Equation (8.1).

$$D_{\text{end-end}} = \sum_{i=1}^{n} d_{\text{prop}} + \sum_{i=1}^{n} d_{\text{que}} + \sum_{i=1}^{n} d_{\text{proc}} + \sum_{i=1}^{n} d_{\text{trans}} \tag{8.1}$$

The time taken for the data to propagate through the network to reach the destination is the propagation time. It is defined as the ratio of the distance (dist) between the neighbour nodes and propagation speed (S_{prop}) of data over the transmission medium and the same is given in Equation (8.2).

$$d_{\text{prop}} = \frac{\text{dist}}{S_{\text{prop}}} \tag{8.2}$$

In general, an IP backbone network is an optical network, and the speed is in the range of 100 Gbps. In such networks, generally, distances between the switches may not vary. The distance between switch$_i$ and switch$_j$ is derived based on the Euclidean distance calculation using Equation (8.3). Here, (x_i, y_i) and (x_j, y_j) are the coordinates of switch$_i$ and switch$_j$, respectively.

$$\text{link\_dist}_{ij} = \sqrt{\left(\left(x_i - x_j \right)^2 + \left(y_i - y_j \right)^2 \right)} \qquad (8.3)$$

The amount of time the data is waiting in the queue at the router/switch is denoted as the queuing delay and it is given in Equation (8.4). It is proportional to the size of the buffer of the switch. Generally, in SDN-based networks, the switch does not make any control decisions as it is done by the controller. Switches only do lookups in the flow table for matching header fields and perform designated actions accordingly. The flow table is stored in TCAM and it provides very fast lookups. Hence, queuing delay in this case is negligible.

$$d_{\text{que}} = \text{waiting\_time\_at\_switch's\_buffer} \qquad (8.4)$$

Processing delay is the amount of time required to process the packet header for a match field in order to forward it accordingly. This is also negligible in the case of normal data traffic. But, in the case of secure data, it is accountable as it involves encryption and decryption. It is given in Equation (8.5).

$$d_{\text{proc}} = \text{time\_to\_examine\_packet\_header} \qquad (8.5)$$

The time taken to push all the bits of the packet onto the medium is the transmission medium. Generally, it is calculated as the ratio between the packet size and bandwidth of the link. It is calculated as given in Equation (8.6)

$$D_{\text{trans}} = \frac{\text{packet\_size}}{\text{bandwidth}} \qquad (8.6)$$

As queuing, processing and transmission delays are with respect to nodes, they are termed a nodal delay. Whereas propagation delay is with respect to the link, it is termed as link delay. Hence, Equation (8.1) can be rewritten as given in Equation (8.7),

$$\text{Delay}(p) = \sum_{e \in E(p)} \text{Delay}(e) + \sum_{n \in N(p)} \text{Delay}(n) \qquad (8.7)$$

Delay(p) represents the delay of a path $p(p \in P)$. It is calculated as the addition of cumulative delay occurred at '$n-1$' edges/links of path p and the cumulative delay occurred at 'n' intermediate nodes of path p. As the switches share common characteristics, there will not be much deviation with respect to nodal delay. Hence, the total delay is highly influenced by the propagation delay (i.e.) link delay. Let D_{max} be the maximum delay tolerable by any application traffic. Then, the path must satisfy the quality requirement as per Equation (8.8).

$$\text{Delay}(p) \leq D_{\text{max}} \qquad (8.8)$$

The paths that are not satisfying the delay quality requirement are exempted from consideration for route calculation.

8.3.3 BANDWIDTH

Bandwidth is defined as the number of bits that can be transferred through the link in a single-time unit usually in one second. As bandwidth is a concave metric, the bandwidth of a path p, Bandwidth(p), is determined as the minimum available bandwidth of all the links in that path. It is illustrated as follows. Figure 8.1 presents the path between switch$_1$ to switch$_4$, and it contains three links namely link$_{1-2}$, link$_{2-3}$ and link$_{3-4}$.

The data rate/bandwidth of link$_{1-2}$, link$_{2-3}$ and link$_{3-4}$ are assigned as 10 Mbps, 5 Mbps and 20 Mbps, respectively. When data traffic enters switch$_1$ and there is no other data flow through link$_{1-2}$, it has 10 Mbps effective bandwidth to reach switch$_2$. But, when the same data traffic flows through link$_{2-3}$, the data rate is reduced because the effective bandwidth is only 5 Mbps. Even though the incoming traffic is with a 10 Mbps data rate, it cannot proceed with the same data rate. Moreover, when this traffic with a 5 Mbps data rate is passed on link$_{3-4}$, the available bandwidth of 20 Mbps could not be utilized fully. Hence, the traffic reaches the destination switch only with a 5 Mbps data rate, and the effective bandwidth of the path is 5 Mbps only. This is the minimum bandwidth of all the links on the path.

Let B_{min} be the minimum bandwidth required for any application traffic, and then the path should satisfy the quality requirement as given in Equation (8.9).

$$\text{Bandwidth}(p) = \left\{ \min\left(\text{Bandwidth}(e)\right), e \in E(p) \right\} \leq B_{min} \qquad (8.9)$$

Bandwidth(e) is the available bandwidth of the edge e belonging to the path p. $E(p)$ contains all edges of the path p. If $B_{min} = 10$ Mbps for an application traffic that enters switch$_1$, the path as given in Figure 8.1 does not satisfy the quality requirement. Hence, this path is rejected by the controller for working or backup path calculation.

8.3.4 CALCULATION OF WORKING AND BACKUP PATHS IN QTR

Let e_{ij} be the link between nodes i and j. Each link in the sample topology is associated with a constraint vector $CV_{ij}[q_1(e_{ij}), q_2(e_{ij})]$. The first quality metric link delay is denoted as $q_1(e_{ij})$ and the second quality metric link bandwidth is denoted as $q_2(e_{ij})$. Let LC(e_{ij}) be the cost of the link. Then, the cost of the link is calculated as given in Equation (8.10).

$$\text{LC}\left(e_{ij}\right) = \alpha \cdot q_1\left(e_{ij}\right) - \beta \cdot q_2\left(e_{ij}\right) \qquad (8.10)$$

The cost of the path is calculated based on the cumulative value of link costs as presented in Equation (8.11). Let $C(p)$ be the cost of the path. Then,

FIGURE 8.1 A sample path from S1 to S4 with available bandwidth.

$$C(p) = \sum_{e_{ij} \in E(p)} \mathrm{LC}(e_{ij}) \qquad (8.11)$$

The cost of the path must meet the quality requirement of the application. Hence, the threshold value for the path cost, say $C(p)_{\mathrm{thresh}}$, is calculated based on the minimum delay and bandwidth requirement of the application. As mentioned previously, $D_{\max}$ is the maximum delay tolerable by an application and $B_{\min}$ is the minimum bandwidth required for that application. Considering these two values, the threshold value for the path cost $C(p)_{\mathrm{thresh}}$ is calculated as per Equation (8.12).

$$C(p)_{\mathrm{thresh}} = \alpha \cdot D_{\max} - \beta \cdot B_{\min} \qquad (8.12)$$

Hence, the cost of the path must not be greater than this threshold value. It is given in Equation (8.13) as follows:

$$C(p) \le C(p)_{\mathrm{thresh}} \qquad (8.13)$$

Generally, interactive multimedia applications have strict end-to-end delay requirements. It should be typically not more than 150–200 ms, whereas normal data traffic accepts a delay even more than this [22]. In case of bandwidth requirement, basic internet applications [23] like web surfing, online games, social media and email require 1–4 Mbps. Mid-range internet applications like single-end video conferencing, online e-commerce and asynchronous online presentations require 3–8 Mbps, whereas high-end internet applications like multi-end video conferencing, telecommuting and distance learning require 20–60 Mbps. The general bandwidth requirements of various internet applications are listed in Table 8.1.

The controller can compute the working and backup paths based on the above calculations. When a new packet_in message comes to the controller, it identifies the APP_TYPE based on the Type of Service field of the packet's header. According to

TABLE 8.1
Bandwidth Requirements of Various Internet Applications

Category of Application Traffic	Required Bandwidth
General web surfing, email, social media	1 Mbps
Online gaming[a]	1–3 Mbps
Video conferencing[b]	1–4 Mbps
Standard-definition video streaming	3–4 Mbps
High-definition video streaming	5–8 Mbps
Frequent large file downloading	50 Mbps and up
Distance learning	50 Mbps and up

[a] A connection with low latency, the time it takes for our computer to talk to the game server, is more important than the bandwidth for gaming.

[b] It requires at least a 1 Mbps upload speed for quality video conferencing.

the predefined quality values for each APP_TYPE, the D_{max} and B_{min} are set. Using this, the threshold value $C(P)_{thresh}$ for the path cost is calculated. The working path that satisfies $C(WP) \leq C(P)_{thresh}$ is installed on the switches' flow tables. This procedure is explained in Figure 8.2. Once the working path entries are done in the

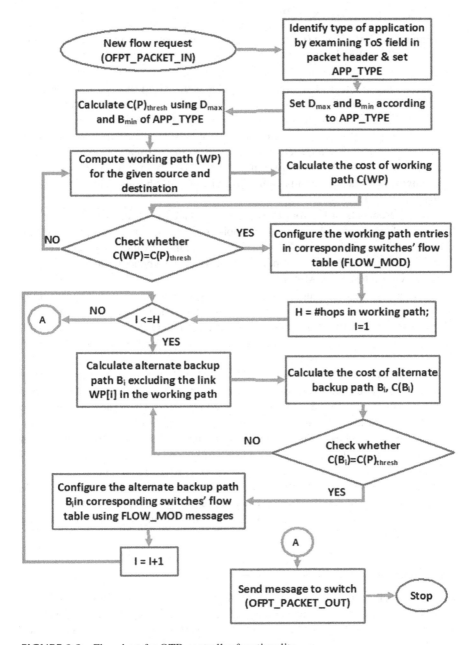

FIGURE 8.2 Flowchart for QTR controller functionality.

switches' flow tables, then the controller computes backup paths for each link on the working path that satisfies the threshold value. Once the backup paths are computed, the corresponding flow rules are installed on the switches' flow tables in order to follow during failures in the working path.

8.4   EXPERIMENTAL SETUP

A Mininet emulator [24–27] is used to experiment with the proposed work. POX controller [28, 29] and Open vSwitch 2.1.0 are used to emulate the SDN environment. Custom network topologies with 5–50 switches and a random number of hosts (between 2 and 4) under each switch are created using MiniEdit. The controllers are placed randomly and the port numbers are configured using controller properties in MiniEdit. The switch configurations are done using MiniEdit preferences.

The network topology is designed in such a way that every networking node in the working path has alternate backup paths to the destination. The OpenFlow controller is modified in order to compute backup paths using the braided multipath algorithm. Each switch is emulated to run Open Shortest Path First routing configured with BFD.

Two different sets of network topologies are created in order to analyse the performance of this work. The first set of topologies is created as tree topologies with varying sizes of switches with different depths and fan-out values. The second set of topologies is created as custom topologies similar to backbone topologies using the MiniEdit GUI interface in Mininet.

Once the topology is set up, the following things are to be verified. The first thing is the port numbers on each switch; it is to be made such that each switch is configured with different port numbers. Then, using 'show OVS summary' command, verify the configurations of switches. To verify the flows in the switches, 'ovs-ofctl dump-flows' command is used.

Xterm and Wireshark are used to monitor the flow. Before running the simulation, start the Command Line Interface (CLI) prompt. In the Xterm window of the source host, a packet trace with the command tcpdump is started. Then, the ping command is run to send traffic between the source and destination host. In the Wireshark window and destination host's Xterm window, that runs tcpdump, it is seen that Internet Control Message Protocol packets are successfully sent and responses are received. Using the 'link-down' command, the broken link is simulated to check for failures. This can be given as a command in CLI or it can be set with the GUI option available in MiniEdit. To do this, right-click on a particular link in the topology and click Link Down. Then continue to send ping commands with varying intervals and it can be monitored for packet loss.

8.5   RESULTS AND DISCUSSION

In order to analyze the performance of QTR, emulation is done with varying values for ping intervals and the varying number of switches. ATR [17] and MLTR [18] are the previous works by the same author. As QTR is an extension to ATR rather than MLTR proposed by the same authors, it is compared only with ATR, and the existing QoS-aware model Hi-QoS is proposed by Jinyao et al. [13]. Among the several existing QoS-aware models, Hi-QoS addresses failure recovery also. Hence, it is appropriate to compare it.

Further, this work focusses on the provisioning of QoS by selecting working and backup paths that satisfy the quality requirement mandated by the application. The link delay and bandwidth are the quality metrics considered for this work because most of the real-time applications require these two. Hence, it is appropriate to compare the performance of this work with respect to average end-to-end delay and bandwidth utilization. As this work is also addressing failure recovery, in addition to average end-to-end delay and bandwidth utilization, average recovery time and PLR are also analysed.

In order to analyse the performance of real-time multimedia data traffic and manipulate the video streaming, in addition to ping traffic generation, a video server is added to stream video contents to a client host. This work considers only normal ping traffic and video traffic only. In future, this may be extended to include voice traffic and interactive gaming also.

8.5.1 ANALYSIS ON AVERAGE END-TO-END DELAY

The emulation is done with a varying number of switches while sending ping traffic as well as streaming a video. Ping packets are sent between a pair of source and destination hosts with an interval of 20 ms. After this, video streaming started between another pair of server and client hosts. The results of both emulations are noted. The results plotted are average values of 15 independent runs.

The results of average end-to-end delays with respect to the varying number of switches are given in Figure 8.3. Since ATR doesn't consider any quality metrics for working and backup path calculation, it results in additional delay than Hi-QoS and QTR. As QTR selects both working and backup paths considering the quality

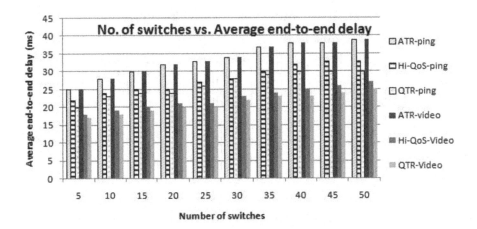

FIGURE 8.3 Number of switches vs. average end-to-end delay.

metrics and chooses only the paths that have sufficient quality resources, it performs better than ATR and Hi-QoS.

From the graph, it is evident that ATR considers ping and video traffic in the same manner. But, Hi-QoS and QTR handle these two applications differently. Hi-QoS achieves this at the cost of maintenance overhead involved in keeping different queues for different types of traffic.

But, QTR achieves this without such an overhead at the networking devices. Though QTR achieves it at the cost of controller overhead in computing QoS-aware paths, it is only algorithmic complexity and satisfies the requirement. It is also seen from the graph that the delay increases with the increase in network size. This may be due to the increase in path length with more nodes in the network. Based on the studies, though the tolerable delay for video traffic is 150–200 ms [22], here, the average end-to-end delay achieved for video traffic is ranging between 15 and 25. This is because of the use of only one video server in the experimentation. But, in real time, there may be multiple video servers streaming simultaneously.

8.5.2 ANALYSIS ON BANDWIDTH UTILIZATION

For the same set of simulations, the results of bandwidth utilization with respect to the varying number of nodes are presented in Figure 8.4.

Since ATR doesn't consider quality metrics for working and backup path calculation, it results in lesser bandwidth utilization than Hi-QoS and QTR. As QTR selects both working and backup paths considering the link delay as well as link bandwidth quality metrics and chooses only the QoS-rich paths, it gives better results than ATR and Hi-QoS. ATR produces the same results for both ping and video traffic whereas Hi-QoS and QTR had produced different results as they handle them accordingly.

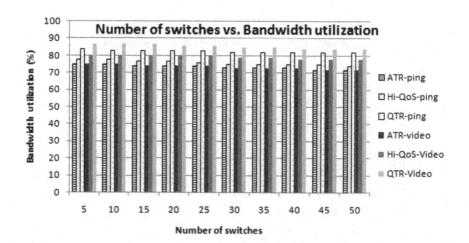

FIGURE 8.4   Number of switches vs. bandwidth utilization.

8.5.3 ANALYSIS ON AVERAGE RECOVERY TIME

Analysis of average recovery time is also done with a varying number of switches. The number of switches varied from 5 to 50 with a constant ping interval of 20 ms. Ping packets are sent between a pair of source and destination hosts. Also, video streaming started between another pair of server and client hosts. As failure recovery is common for any type of traffic, this part considers both the traffic commonly. This emulation is repeated for 15 runs in order to calculate the average value. In each run, different links one at a time are made to fail using the 'link-down' command in order to simulate the link failure. The results of node-disjoint multipath, ATR, Hi-QoS and QTR with respect to average recovery time are plotted in the graph presented in Figure 8.5.

From the graph, it is seen that ATR, Hi-QoS and QTR have similar performances with respect to the recovery time but the node-disjoint algorithm has little higher recovery time. When the number of switches increases, the average recovery time also increases slightly. This may be resulted due to the long path length as the number of switches in the topology is increasing. However, it is less than the tolerable standard recovery time of 50 ms [30] while using the proactive protection-based method for failure recovery. Although ATR, Hi-QoS and QTR seem like giving almost similar results, QTR resulted in a 2%–7% improvement with respect to ATR and 2%–4% with respect to Hi-QoS.

8.5.4 ANALYSIS ON PACKET LOSS RATIO

This analysis is also done with the varying number of switches. With the number of switches ranging from 5 to 50 and a constant ping interval of 20 ms, the results on

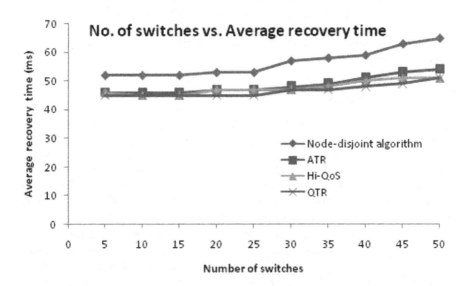

FIGURE 8.5 Number of switches vs. average recovery time.

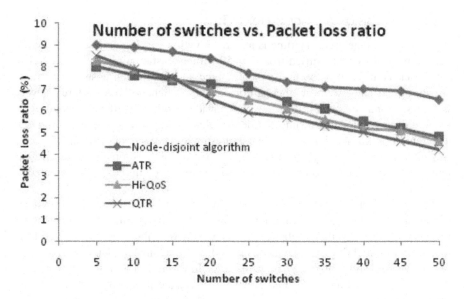

FIGURE 8.6   Number of switches vs. packet loss ratio.

PLR are depicted in Figure 8.6. Also, video streaming started between another pair of server and client hosts. As this measure is common for any type of traffic, this part considers both the traffic commonly. From the graph, it is noticed that the PLR decreases with the increase in network size. This may be due to the high availability of alternate paths through different switches. Initially, up to 15 nodes, QTR showed a higher PLR, and after that, it is reduced. This may be due to the unavailability of QoS-rich paths with a short network size. QTR has produced up to 15% improvement over ATR and up to 9% improvement over Hi-QoS. As the node-disjoint algorithm doesn't consider any specific failure recovery mechanism, it has little higher PLR compared to others.

8.6   CONCLUSION

Being a hot research in traditional networking architecture as well as in today's SDN architecture, provisioning of QoS became the mandatory requirement for real-time internet applications. It needs to handle different traffic requirements differently. With the advent of the SDN paradigm and OpenFlow protocol, this had been made easier and flexible to achieve. This work proposed an effective scheme towards QTR and also served as an effective method for failure recovery in an OpenFlow-based SDN. This work chooses working and backup paths based on the link delay and link bandwidth metrics. The paths that satisfy the minimum delay and bandwidth requirements for specific applications are only considered for routing. Experiments showed that this work produced better results with respect to average end-to-end delay and bandwidth utilization. In future, this work may be extended to consider control plane failures also using a multi-controller environment in addition to forward plane failures.

REFERENCES

1. I.F. Akyildiz, A. Lee, P. Wang, M. Luo, & W. Chou, Research challenges for traffic engineering in software defined networks, *IEEE Network*, Vol. 30, No. 3, pp. 52–58, May-June, 2016.
2. D. Turner, K. Levchenko, A.C. Snoeren, & S. Savage, California fault lines: Understanding the causes and impact of network failures, *ACM SIGCOMM Computer Communication Review*, Vol. 41, No. 4, New York, United States, August 2011, pp. 315–326.
3. M. Karakus, & A. Durresi, Quality of Service (QoS) in Software Defined Networking (SDN): A survey, *Journal of Network and Computer Applications*, Vol. 80, pp. 200–218, February, 2017.
4. G. Muthusamy, & S.R. Chandran, Cluster-based task scheduling using K-means clustering for load balancing in cloud datacenters, *Journal of Internet Technology*, Vol. 22, No. 1, pp. 121–130, January, 2021.
5. D. Katz, & D. Ward, Bidirectional forwarding detection (BFD), (No. RFC 5880). https://tools.ietf.org/html/rfc5880#section-4.1, June, 2010.
6. S. Civanlar, M. Parlakisik, A.M. Tekalp, B. Gorkemli, B. Kaytaz, & E. Onem, A QoS-enabled OpenFlow environment for scalable video streaming. *IEEE GLOBECOM Workshops*, Miami, Florida, USA, December 2010, pp. 351–356.
7. H. Kim, M. Schlansker, J.R. Santos, J. Tourrilhes, Y. Turner, & N. Feamster, CORONET: Fault tolerance for software defined networks. *20th IEEE International Conference on Network Protocols (ICNP)*, Austin, Texas, USA, October–November 2012, pp. 1–2.
8. H.E. Egilmez, S.T. Dane, K.T. Bagci, & A.M. Tekalp, OpenQoS: An OpenFlow controller design for multimedia delivery with end-to-end Quality of Service over Software-Defined Networks. *IEEE Signal & Information processing association annual summit and conference (APSIPA ASC)*, Asia-Pacific, Hollywood, CA, USA, December 2012, pp. 1–8.
9. H.E. Egilmez, S. Civanlar, & A.M. Tekalp, An optimization framework for QoS-enabled adaptive video streaming over OpenFlow networks, *IEEE Transactions on Multimedia*, Vol. 15, No.3, pp. 710–715, April, 2013.
10. H.E. Egilmez, & A.M. Tekalp, Distributed QoS architectures for multimedia streaming over software defined networks, *IEEE Transactions on Multimedia*, Vol. 16, No. 6, pp. 1597–1609, October, 2014.
11. W. Wendong, Q. Qinglei, G. Xiangyang, H. Yannan, & Q. Xirong, Autonomic QoS management mechanism in software defined network, *China Communications*, Vol. 11, No. 7, pp. 13–23, July, 2014.
12. K. Govindarajan, K.C. Meng, H. Ong, W.M. Tat, S. Sivanand, & L.S. Leong, Realizing the quality of service (QoS) in software-defined networking (SDN) based cloud infrastructure. *2nd IEEE International Conference on In Information and Communication Technology (ICoICT)*, Bandung, Indonesia, May 2014, pp. 505–510.
13. Y. Jinyao, Z. Hailong, S. Qianjun, L. Bo, & G. Xiao, HiQoS: An SDN-based multipath QoS solution, *China Communications*, Vol. 12, No. 5, pp. 123–133, May, 2015.
14. G. Xu, B. Dai, B. Huang, J. Yang, & S. Wen, Bandwidth-aware energy efficient flow scheduling with SDN in data center networks, *Future Generation Computer Systems*, Vol. 68, pp. 163–174, March, 2017.
15. W. Zhang, X. Zhang, H. Shi, & L. Zhou, An efficient latency monitoring scheme in software defined networks, *Future Generation Computer Systems*, Vol. 83, pp. 303–309, June, 2018.
16. M. Bagaa, D.LC. Dutra, T. Taleb, & K. Samdanis, On SDN-driven network optimization and QoS aware routing using multiple paths, *IEEE Transactions on Wireless Communications*, Vol. 19, No.7, pp. 4700–4714, July, 2020.

17. N. Senthilkumaran, & R. Thangarajan, Habitual failure recovery by traffic rerouting with pre-configured backup paths in OpenFlow based software defined networks, *Journal of Web Engineering*, Vol. 17, No. 6, pp. 3320–3338, May, 2018.

18. N. Senthilkumaran, R. Thangarajan, & S.K. Nivetha, Memory and Load-aware Traffic Rerouting (MLTR) in OpenFlow-based SDN. *TEQIP III Sponsored IEEE International Conference on Microwave Integrated Circuits, Photonics and Wireless Networks (IMICPW'19)*, Trichy, Tamilnadu, May 2019, pp. 409–413.

19. L. Hanzo, & R. Tafazolli, A survey of QoS routing solutions for mobile ad hoc networks, *IEEE Communications Surveys & Tutorials*, Vol. 9, No. 2, pp. 50–70, July, 2007.

20. T. Kenyon, *High Performance Data Network Design: Design Techniques and Tools*, Digital Press, Oxford, pp. 185–247, 2002.

21. A. Sheikh, & A. Ambhaikar, Quality of services parameters for architectural patterns of IoT, *Journal of Information Technology Management*, Special Issue, 36–53, 2021.

22. J.D. Saunders, C.R. McClure, & L.H. Mandel, Broadband applications: Categories, requirements, and future frameworks, *First Monday*, Vol. 17, No. 11, November, 2012.

23. https://www.nerdwallet.com/blog/utilities/how-to-choose-the-best-internet-service/

24. K. Kaur, J. Singh, & N.S. Ghumman, Mininet as software defined networking testing platform. *International Conference on Communication, Computing & Systems (ICCCS)*, Ferozepur, Punjab, India, August 2014, pp. 139–142.

25. R.L.S. De Oliveira, A.A. Shinoda, C.M. Schweitzer, & L.R. Prete, Using mininet for emulation and prototyping software-defined networks. *IEEE Colombian Conference on Communications and Computing (COLCOM)*, Bogotá, D.C., Colombia, June 2014, pp. 1–6.

26. F. Keti, & S. Askar, Emulation of software defined networks using mininet in different simulation environments. *6th International Conference on Intelligent Systems, Modelling and Simulation*. IEEE Computer Society, NW Washington, DC, United States, February 2015, pp 205–210.

27. C. DeCusatis, A. Carranza, & J. Delgado-Caceres, Modeling software defined networks using mininet. *2nd International Conference on Computer and Information Science and Technology (CIST'16)*, Ottawa, Canada, May 2016, No. 133, pp. 1–6.

28. POX at https://openflow.stanford.edu/display/ONL/ POX+Wiki.

29. K. Kaur, J. Singh, & N.S. Ghumman, Network programmability using POX controller. *International Conference on Communication, Computing & Systems (ICCCS)*, Ferozepur, Punjab, India, August 2014, pp. 134–138.

30. J. Ali, G.M. Lee, B.H. Roh, D.K. Ryu, & G. Park, Software-defined networking approaches for link failure recovery: A survey. *Sustainability*, Vol. 12, No. 10, pp. 4255, May, 2020.

9 Application-Based Evaluation of Neural Network Architectures for Edge Devices

Amit Mankodi, Het Suthar, and Amit Bhatt
Dirubhai Ambani Institute of Information
and Communication Technology

CONTENTS

9.1 INTRODUCTION

In the era of Internet-of-Things (IoT), there is a significant increase in smart networked devices from drones to home appliances and many more. According to Ref. [1], 14.6 billion IoT devices will be deployed by 2022. Therefore, the data generated or collected from these devices in the form of video, audio, text, and sensor data is escalating [2]. Furthermore, due to the advancements in the compute and storage capabilities, utilization of neural network-based machine learning models has increased in applications such as vision sensing (image processing) from images or videos, speech recognition from audio [3,4], and many other data analytic applications.

Until recently, applications involving neural network models were deployed on cloud servers due to the large capacities of computation and storage they offer. However, this requires a large amount of data to be transported over the network from the source of the data to the cloud servers requiring large network bandwidth and adding latency in processing. Additionally, cloud servers have large heat profiles due to significant energy consumption causing environmental hazards.

Many applications such as cognitive or real-time applications require faster response times and lower energy consumption which has led to the development of edge computing [5]. Edge computing enables deploying neural network models for the said applications to process the data or perform data analytics nearby the source of the data [2] on low-power embedded systems [6]. Edge computing is possible today due to the technological advancements in microcontroller-based embedded systems providing enhanced computation and storage capacities. These devices are packed with either high-performance processors such as ARM Cortex-A or Cortex-R or may have ultra-low-power microcontrollers such as Cortex-M [6].

Two approaches have become popular to implement edge computing. In the first approach, both training and inference are performed on edge devices [7–9]. The second approach is to train the model on servers and then convert the model to reduce the memory footprint to fit on the microcontroller systems for inference. The second approach has gained significant research interest in the industry [10–12] and academia [2] due to its wide area of applications (TinyML) [13], and therefore, is the primary aspect of our work.

Evaluating the performance of neural network models with different architectures on microcontroller systems with dissimilar computing and storage capacities for a given application is an important step for benchmarking [14]. Therefore, in this work, we first developed seven neural network models with dissimilar architectures on a cloud server for three applications: vision sensing, keyword detection, and gesture recognition. We applied quantization to the trained model to fit the model in a small memory and then deployed the quantized model on three microcontroller systems. We measured latency and energy consumption during the inference on all the microcontroller systems for evaluation. The quantized models achieved an accuracy of at least 80% in most cases. Our results show that a microcontroller system with a Cortex-M4F processor has the lowest energy consumption for all three applications. However, vision sensing and keyword detection have the lowest latency for the Cortex-M4 system while gesture detection has the lowest latency for the Cortex-M4F system.

9.2   NEURAL NETWORK MODELS AND MICROCONTROLLER SYSTEMS

In this section, we provide details about neural network models and embedded systems with microcontrollers used for the experiments. We also provide a detailed process of quantization that allows us to use neural network models on low-power devices with small memory and compute footprints.

9.2.1   NEURAL NETWORK ARCHITECTURES

singleFC: The singleFC architecture is one of the simplest neural networks. The input data is reshaped to feed directly to a fully connected layer followed by a SoftMax activation function $\sigma(z_i) = \dfrac{e^{z_i}}{\sum\limits_{j=1}^{K} e^{z_j}}$ that classifies the input into one of the several

classification categories. This model requires a small amount of memory and fewer computations compared to the other neural network models described in this section.

CNN: Convolutional neural network (CNN) architecture typically has three layers: convolutional, pooling, and fully connected layer. A convolutional layer takes data (e.g., an image) as an input and applies filters (kernels) to the image to extract correlational features $(i * f)[m,n] = \sum\limits_{j}\sum\limits_{k} f[j,k] * i[m-j, n-k]$ with i and f representing image and filter matrices. The general operating principle of a convolutional layer is shown in Figure 9.1. The values of these filters are decided by backpropagation. The length, height, and number of filters determine the length, height, and depth of the subsequent layer. The convolutional layer is followed by a pooling layer with a ReLU activation function (max pooling) to reduce the dimensions of the extracted features. Lastly, a fully connected layer is connected to the SoftMax activation function for producing the desired classification output.

CNN-2: CNN-2 architecture is similar to CNN except it only consists of one depth-wise convolutional layer followed by a fully connected layer with a SoftMax activation to obtain the output. This architecture requires larger memory size and computations than a singleFC.

DNN: A deep neural network (DNN) is made up of multiple CNN layers with ReLU activation and max-pooling layers added in between the convolutional layers to reduce the feature dimensions and increase the trainable parameters. The DNN architecture consists of two hidden fully connected layers with ReLU and SoftMax activation, respectively. The layers with ReLU activation function determine the output using $z = \sum\limits_{l=1}^{n_1} W_l * \left(\sum\limits_{k=1}^{n_2} W_{kl} \dots * \left(\sum\limits_{j=1}^{n_{m-1}} (W_{jk}) * \left(\sum\limits_{i=1}^{n_m} W_{ij} * X_i \right) \right) \right)$ while the last layer with SoftMax activation function derives the output using $\sigma(z_i) = \dfrac{e^{z_i}}{\sum\limits_{j=1}^{K} e^{z_j}}$.

MobileNet: The shallow MobileNet [15] architecture is modified to support a grayscale 32×32 input image and a depth multiplier of 0.25. MobileNet uses a

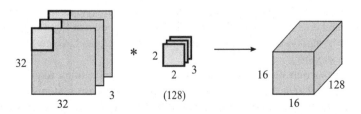

FIGURE 9.1   Operating principle of convolution.

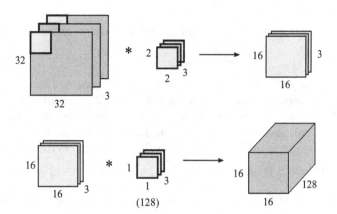

FIGURE 9.2 Operating principle of depth-wise separable convolution.

depth-wise separable convolution instead of the regular convolution operation. A depth-wise separable convolution consists of two operations: a 2D convolution with an individual channel of input followed by a point-wise convolution (i.e., 1×1) that determines the depth of the output, as shown in Figure 9.2. The depth-wise separable convolution proves to be efficient in terms of the number of operations as well as parameters.

DS-CNN: Inspired by the MobileNet architecture, depthwise separable convolutional neural network (DS-CNN) architecture was developed to replace the two convolutional layers of DNN with a depth-wise separable convolution to observe how the DNN would perform if its convolutional layers were replaced by depth-wise separable layers.

DS-CNN-2: DS-CNN-2 architecture is a modified version of DS-CNN to have a greater number of trainable parameters by changing the convolutional kernel size and stride value. Although the model has 1.2 times more trainable parameters, the features extracted by the architecture are not optimum.

9.2.2 MICROCONTROLLER SYSTEMS

Microcontroller systems usually consist of a processor operating at a few MHz of frequency, limited on-chip memory of a few kilobytes, and an on-chip flash. Table 9.1 shows the embedded boards with dissimilar Cortex-M processors and memory configurations. The Cortex-M4 processor cores have a 3-stage pipeline as shown in Figure 9.3a. On the other hand, Cortex-M7 processor cores have a 6-stage superscalar pipeline with instruction and data cache of 4KB each; the pipeline is shown in Figure 9.3b. Both M4 and M7 cores support the Thumb/Thumb2 instruction set and have an 8/16-bit SIMD architecture support.

These microcontroller systems are designed for applications requiring a small amount of power (energy) consumption. Furthermore, these systems generally do not have adequate resources for compute-intensive tasks such as neural networks.

TABLE 9.1

Microcontroller-Based Embedded Board Systems

Board Name	Microcontroller	Frequency	SRAM	Flash
Arduino Nano 33 BLE Sense	Cortex-M4	64 MHz	256 KB	1 MB
SparkFun Edge Apollo 3 Blue	Cortex-M4F	48 MHz	384 KB	1 MB
STM32F746NG	Cortex-M7	216 MHz	320 KB	1 MB

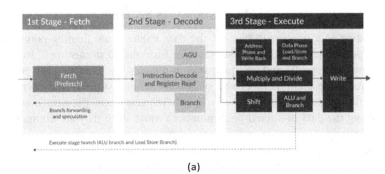

(a)

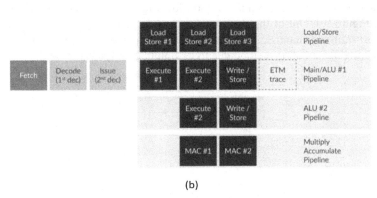

(b)

FIGURE 9.3   Cortex-M pipeline stages. (a) Cortex-M4 [16], (b) Cortex-M7 [17].

However, some architecture such as Cortex-M4 and Cortex-M7 have support for SIMD and MAC instructions which can be used to perform efficient computations of neural networks. Furthermore, the ability to perform neural network inference on microcontroller systems without any external supporting hardware is enabled by TinyML [13]. In a joint effort from Google and ARM, CMSIS-NN [11] kernel was developed for optimized implementation of various neural network functions on Cortex-M-based microcontrollers, and required libraries were provided in TensorFlow to convert and save the weights of a trained model.

9.3   DEPLOYING NEURAL NETWORK MODEL
ON MICROCONTROLLER SYSTEMS

Deploying a neural network requires mainly three tasks: data collection and preparation of a dataset, training a neural network on the collected dataset, and finally performing inference/prediction using the trained neural network. The data values used as input and during the neural network training are represented by floating-point numbers. However, floating-point operations are computationally more expensive and require larger memory for storage. Hence, the solution is to convert these data values from floating-point (real) numbers to fixed-point (integer) numbers.

9.3.1   QUANTIZATION FOR MICROCONTROLLER SYSTEMS

Quantization is a process to convert the data values from floating-point numbers to fixed-point integers reducing them in size for reduced storage and also enabling faster computations. Therefore, quantization reduces the storage space required to fit the data in the small memory size of microcontroller systems. Furthermore, low-power microcontroller systems with lower frequency can perform computations faster using the data in a fixed-point number format since they are computationally less expensive.

To deploy a neural network on microcontroller systems, quantization needs to be applied to the weights and biases of the neural network. For this, we used TensorFlow [18] Lite Converter utility from the TensorFlow Lite library as shown in the process flow in Figure 9.4. Out of the two quantization methods supported by TensorFlow Lite, post-training quantization, and quantization-aware training, we utilized the post-training quantization method due to its simplicity.

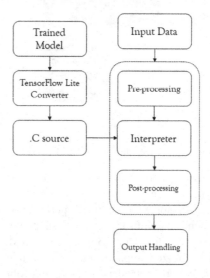

FIGURE 9.4   Workflow for deploying neural networks on microcontroller systems.

We first trained the model using float32 numbers on Google Colaboratory. We then applied post-training quantization to the trained model. This method is easy to apply but comes with a reduction in accuracy. By converting from float32 to float16, one can save 50% of space in memory, with little or no loss in accuracy. However, microcontroller systems used in this work have very limited memory and computing capabilities; hence, we quantized the models from float32 to int8, saving up to 75% in memory and faster computation.

9.3.2   Using Quantized Neural Network Model in Application

Including Dependencies: The application code is written in C++, and it uses many library functions from the TensorFlow Lite library. First, we identify the dependencies required to be included in the code for the model to work. These header files include most of the codes required by the TensorFlow Lite framework to be able to perform inference on a quantized neural network model and produce an output.

```
#include "tensorflow/lite/micro/examples/hello_world/model_
quant.h" // Consist of the model that was trained and
converted to C++ using xxd.
#include "tensorflow/lite/micro/kernels/all_ops_resolver.h" //
A Class that loads the operations used by our neural network
model.
#include "tensorflow/lite/micro/micro_error_reporter.h"
#include "tensorflow/lite/micro/micro_interpreter.h" //
TensorFlow Lite interpreter, that will run the t rained
model.
#include "tensorflow/lite/schema/schema_generated.h" // Schema
that defines the structure of the data stored in model\quant
.h file
#include "tensorflow/lite/version.h"
```

Loading the Model: The next step is to map the model to the application code. The code below creates a pointer to the *model* with the name of the *model*, which represents the neural network model to be used.

```
const tflite :: Model* model = :: tflite ::
GetModel(g_sine_model_data);
```

Now, we need to load the model operations to perform inference using the loaded model. We can either load all the available operations using a single statement as shown below or we can only load the selected operations such as SoftMax, ReLU, or MaxPooling required by the model.

```
// Pull in all the operation implementations
tflite :: ops :: micro :: AllOpsResolver resolver;
```

Defining a Tensor Arena: Now that we have loaded our model and the required operations, we need to allocate an area of memory that will be used by our model when it executes.

```
const int tensor_arena_size = 2 * 1024;
uint8_t tensor arena[tensor_arena_size];
```

The inputs to the model, intermediate results, and outputs will be stored in the memory. It is in the multiples of 1,024 bytes, the size of the tensor arena depends on the application and size of the model, and there is no fixed number that always works. The only limitation is that the memory requirement needs to be small enough to fit in the available microcontroller system's SRAM.

Performing Inference: To perform inference, we need to provide input to the interpreter, which can be done by writing the input data to the model's input tensor. TensorFlow also provides a variable called "data" of union type, which can store the input data.

```
TfLiteTensor* input = interpreter.input(0);
input->data.f[0] = 0.;
```

Now that the input data is provided, we need to run our model, i.e., perform inference on the input data and read the output. The first step is to invoke the interpreter; this will perform inference on the input data and write the results in the output tensor. Similar to the input data, output also is written in a union variable "data" which can be read to know the result.

```
TfLiteStatus invoke_status = interpreter.Invoke();
TfLiteTensor* output = interpreter.output(0);
float value = output->data[0];
```

Once the output is available, it can be used on different microcontroller systems to perform the required functionality. The basic framework of performing inference remains the same across all microcontroller systems, except for two changes: first, how the input is obtained from different hardware peripherals, and the other, how the output is processed.

9.4   EXPERIMENTS AND RESULTS

In this section, we consider fire detection a vision-sensing application, keyword detection from audio, and gesture recognition application to analyze the configurations of neural network models detailed in Section 2.1. We selected a different set of neural network architectures for the three applications according to the shape and characteristics of the input data. The neural network models were first constructed on Google Colaboratory and then deployed on microcontroller systems listed in Section 2.2 by following the procedure detailed in Section 9.3. To measure latency, we sent a high value (1) to one of the I/O pins of microcontroller systems when the inference started and a low value (0) when the inference finished and measured the duration (latency) of high value on the digital oscilloscope. To measure energy consumption, we utilized a source meter to first measure voltage and current while performing inference, and then the energy consumption was calculated using the current consumption, voltage, and latency (energy consumed = current consumption × voltage × latency).

9.4.1 Vision Sensing

Vision-sensing applications can be seen everywhere, from smart security cameras to self-driving cars and building surveillance systems. In a general video surveillance system, the object of interest is not present in the video stream most of the time. But the video stream is sent from a camera to a cloud server continuously for the detection of a specific object, which requires transmission of the data with higher bandwidth. Furthermore, cloud servers consume more power than microcontroller systems while performing object detection.

The microcontroller systems, therefore, make a good contender for vision sensing applications on the edge, as these are low-power systems and economical to deploy on large scale. Microcontroller systems can be always on, processing the video stream on-device and only setting a trigger to a larger cloud system upon detecting the object of interest. This type of application is also known as "Visual Wake Word [19]." Because either the microcontroller systems are very close to the camera or the camera of these systems is used for capturing the video, there is no need to send the video stream to the cloud. Additionally, low-power microcontroller systems consume much less power compared to cloud servers for object detection.

We used the fire image dataset [20] to train the neural network architectures for different input sizes viz. $32 \times 32 \times 3$ for red, green, and blue (RGB) and $32 \times 32 \times 1$ for grayscale. Training was performed on Google Colaboratory for over 900 images with 1,000 epochs, and then the trained models were quantized. The memory requirements and prediction accuracies were measured before and after the quantization as depicted in Tables 9.2 and 9.3. It is observed that DS-CNN provides better accuracy for RGB images due to the depth of three, but grayscale images do not benefit from a depth of one. However, DS-CNN requires higher million operations (MOps) per inference compared to MobileNet due to the additional convolution for the depth-wise layer.

TABLE 9.2

Fire Detection (Input Image in RGB $32 \times 32 \times 3$): Neural Network Memory Requirements and Prediction Accuracy

Neural Network	Accuracy (before Quant)	Memory (before Quant)	Accuracy (after Quant)	Memory (after Quant)	MOps/ Inference
CNN	92%	186 KB	82%	26.77 KB	3.6
DNN	94%	1.6 MB	54%	207.16 KB	5
MobileNet	96%	1.3 MB	32%	181.97 KB	2.92
DS-CNN	76%	714 KB	50%	311 KB	11.82

TABLE 9.3

Fire Detection (Input Image in Grayscale 32×32×1): Neural Network Memory Requirements and Prediction Accuracy

Neural Network	Accuracy (before Quant)	Memory (before Quant)	Accuracy (after Quant)	Memory (after Quant)	MOps/ Inference
CNN	86%	170 KB	80%	24.77 KB	2.62
DNN	94%	1.6 MB	84%	206.59 KB	4.4
MobileNet	92%	1.3 MB	84%	181.83 KB	2.88
DS-CNN	94%	711 KB	80%	109.63 KB	11.67

TABLE 9.4

Fire Detection (Input Image in RGB 32×32×3): Latency and Energy Consumption from Microcontrollers

	CNN	DNN	MobileNet	DS-CNN
		Arduino Nano BLE 33 Sense		
Latency (ms)	375	484	126	180
Energy (mJ)	22.89	29.54	7.69	10.98
		STM32F764NG		
Latency (ms)	1528	1912	213	792
Energy (mJ)	3820	4780	532.5	1980

TABLE 9.5

Fire Detection (Input Image in Grayscale 32×32×1): Latency and Energy Consumption from Microcontrollers

	CNN	DNN	MobileNet	DS-CNN
		Arduino Nano BLE 33 Sense		
Latency (ms)	295	432	122	167
Energy (mJ)	18	26.37	7.44	10.19
		SparkFun Edge		
Latency (ms)	720	860	137	376
Energy (mJ)	12.59	15.04	2.39	6.57
		STM32F764NG		
Latency (ms)	1122	370	198	734
Energy (mJ)	2805	925	495	1835

The quantized trained neural network models were then implemented on micro-controller systems, and latencies and energy consumed were measured as shown in Tables 9.4 and 9.5. We did not include energy consumed by peripherals such as camera modules and onboard LEDs in the measurement since we are only interested in the neural network model's energy consumption. The results indicate that architectures with depth-wise separable convolutional layers, MobileNet and DS-CNN, perform better for microcontroller systems in terms of latencies, MOps, and energy consumptions. Furthermore, Arduino Nano BLE 33 Sense with Contex-M4 has lower latencies whereas SparkFun Edge with Cortex-M4F has lower power consumption.

9.4.2 Keyword Detection

Similar to vision sensing, keyword detection also requires an always-on device to be actively listening for a trigger word. Hence, microcontroller systems are the preferred choice for keyword detection as well rather than sending the audio signal to servers for processing. The input audio signal is first converted into a 2D spectrogram of 49×40 pixels, instead of taking an entire one-second audio sample. We took 49 pieces of 30-ms audio segments with a stride of 20 ms and converted the segments into the frequency domain by taking the fast Fourier transform. This 2D spectrum is then used as an input to the neural networks.

For keyword detection, the speech command dataset [21] was used for training the neural network models. Table 9.6 shows before and after quantization memory requirements and accuracies plus the MOps/inference for each of the neural networks used for keyword detection.

After training the models, the quantized models were deployed on the microcontroller systems. The latency and energy consumption per inference of these models are shown in Table 9.7. It is observed that the singleFC architecture consumes less energy and has a faster response time, but the model is not very accurate whereas the CNN architecture provides good accuracy, but it consumes a lot of space and energy and its response time is not efficient either. The CNN-2 architecture lies in between the singleFC and CNN architectures, providing an accuracy close to that of CNN while performing 13× faster inference and consuming 16.4× less memory. Furthermore, Arduino Nano BLE 33 Sense with Cortex-M4 has lower latency whereas SparkFun Edge with Cortex-M4F has lower energy consumption for CNN variants.

9.4.3 Gesture Recognition

The gesture recognition system uses an accelerometer to measure the acceleration of the sensor in three dimensions and hence the motion. The output of an accelerometer

---

TABLE 9.6

Keyword Spotting: Neural Network Memory Requirements and Prediction Accuracy

Neural Network	Accuracy (before Quant)	Memory (before Quant)	Accuracy (after Quant)	Memory (after Quant)	MOps/ Inference
CNN	95.4%	2.05 MB	94.9%	300.94 KB	101.99
CNN-2	90.9%	137 KB	90.1%	18.27 KB	0.33
SingleFC	82.0%	59.75 KB	81.63%	8.66 KB	0.00078

TABLE 9.7

Keyword Spotting: Latency and Energy Consumption from Microcontrollers

	CNN	SingleFC	CNN-2
	Arduino Nano BLE 33 Sense		
Latency (ms)	10137	0.7	68.78
Energy (mJ)	618.86	0.042	4.19
	SparkFun Edge		
Latency (ms)	19600	2	150
Energy (mJ)	342.80	0.034	128.62
	STM32F764NG		
Latency (ms)	35.7e3	0.478	150
Energy (mJ)	89250	1.19	321.55

TABLE 9.8

Gesture Recognition: Neural Network Memory Requirements and Prediction Accuracy

Neural Network	Accuracy (before Quant)	Memory (before Quant)	Accuracy (after Quant)	Memory (after Quant)	MOps/ Inference
CNN	93.4%	90 KB	93.1%	19.8 KB	0.0062
DS-CNN	96.35%	88.4 KB	95.83%	20.08 KB	0.36
DS-CNN-2	85.62%	134.24 KB	85%	24.86 KB	0.44
DNN	94.79%	0.024 KB	93.68%	0.0049 KB	13.07

consists of three values indicating acceleration in each direction. The neural network architectures have been trained to take a set of 128 such values and then try to predict the gesture. The gesture recognition dataset [13] consists of multiple samples of three different gestures viz. "W", "O", and "L" were recorded by four different people.

A three-axis accelerometer is available on the Arduino BLE 33 Sense and SparkFun Edge boards; the STM32F746NG board doesn't have an on-board accelerometer and hence will not be featured in this study. The SparkFun board can read the 3-axis data at a rate of 25 Hz; this data is directly passed to the neural network without any preprocessing.

Table 9.8 displays memory requirements and accuracies before and after the quantization of models along with MOps/inference. The latencies and energy consumptions for Arduino BLE 33 Sense and SparkFun Edge boards are shown in Table 9.9. It can be observed that these models require a smaller number of MOps as compared to the previous two applications and can achieve very high accuracy with a small memory footprint. After deploying these models on a set of microcontroller systems, it was observed that depth-wise separable convolution architectures, DS-CNN and DS-CNN-2, have lower latencies and energy consumptions similar to vision applications. Furthermore, SparkFun Edge with Cortex-M4F has the lowest energy consumption and latency for all neural network architectures.

TABLE 9.9
Gesture Recognition: Latency and Energy Consumption from Microcontrollers

	CNN	DS-CNN	DS-CNN-2	DNN
	Arduino Nano BLE 33 Sense			
Latency (ms)	160	52.3	54.16	124
Energy (mJ)	9.76	3.19	3.30	7.57
	SparkFun Edge			
Latency (ms)	31.6	10.8	16	28.8
Energy (mJ)	0.55	0.18	0.50	0.27

9.5 CONCLUSIONS

In this work, we have demonstrated that neural network models can be deployed on microcontroller systems to perform compute-intensive tasks such as fire detection, gesture recognition, or keyword detection. With the help of quantization, we have deployed neural network models with large memory footprints to a very small memory space of microcontroller systems. We performed quantization using the TensorFlow Lite Converter which converts floating-point numbers to fixed-point numbers for reducing the memory footprint of the neural network model. We have evaluated three to four neural network architectures for each of the three applications: vision sensing, keyword detection, and gesture recognition. The results demonstrated that depth-wise separable neural network architecture performs better for visual applications such as vision sensing and gesture recognition; however, simple audio application for keyword detection can be performed effectively using a simple neural network model such as singleFC or CNN. Furthermore, SparkFun Edge with Cortex-M4F has the lowest energy consumption so can be chosen where power consumption is the constraint. On the other hand, in the case of latency constraints, Arduino Nano BLE 33 Sense with Cortex-M4 or SparkFun Edge with Cortex-M4F can be selected.

REFERENCES

1. Reports, C.A.: Cisco networking trends. *Cisco* [Online] **5** (2020), https://info.data3. com/l/575413/2020-08-04/6p27vr/575413/127058/2020globalnetworkingtrendsreportp teen_v4.pdf.
2. Shi, W., Cao, J., Zhang, Q., Li, Y., Xu, L.: Edge computing: Vision and challenges. *IEEE Internet of Things Journal* **3**(5), 637–646 (2016). https://doi.org/10.1109/ JIOT.2016.2579198.
3. Armbrust, M., Fox, A., Griffith, R., Joseph, A.D., Katz, R., Konwinski, A., Lee, G., Patterson, D., Rabkin, A., Stoica, I., Zaharia, M.: A view of cloud computing. *Communications of the ACM* **53**(4), 50–58 (2010). https://doi. org/10.1145/1721654.1721672.
4. Dean, J.: The deep learning revolution and its implications for computer architecture and chip design. CoRR **abs/1911.05289** (2019), http://arxiv.org/abs/1911.05289.
5. Satyanarayanan, M., Davies, N.: Augmenting cognition through edge computing. *Computer* **52**(7), 37–46 (2019). https://doi.org/10.1109/MC.2019.2911878.
6. Sanchez-Iborra, R., Skarmeta, A.F.: Tinyml-enabled frugal smart objects: Challenges and opportunities. *IEEE Circuits and Systems Magazine* **20**(3), 4–18 (2020). https://doi. org/10.1109/MCAS.2020.3005467.

7. Alnemari, M., Bagherzadeh, N.: Efficient deep neural networks for edge computing. In: *2019 IEEE International Conference on Edge Computing (EDGE)*. pp. 1–7 (2019). https://doi.org/10.1109/EDGE.2019.00014.

8. Dean, J., Patterson, D., Young, C.: A new golden age in computer architecture: Empowering the machine-learning revolution. *IEEE Micro* **38**(2), 21–29 (2018). https://doi.org/10.1109/MM.2018.112130030.

9. Jouppi, N., Young, C., Patil, N., Patterson, D.: Motivation for and evaluation of the first tensor processing unit. *IEEE Micro* **38**(3), 10–19 (2018). https://doi.org/10.1109/MM.2018.032271057.

10. Doyu, H., Morabito, R., Ḧoller, J.: Bringing machine learning to the deepest IOT edge with TinyML as-a-service (2020), https://iot.ieee.org/newsletter/march-2020/bringing-machine-learning-to-the-deepest-iot-edge-with-tinyml-as-a-service.

11. Lai, L., Suda, N., Chandra, V.: CMSIS-NN: efficient neural network kernels for arm cortex-M CPUS. CoRR **abs/1801.06601** (2018), http://arxiv.org/abs/1801.06601.

12. Zhang, Y., Suda, N., Lai, L., Chandra, V.: Hello edge: Keyword spotting on micro-controllers. CoRR **abs/1711.07128** (2017), http://arxiv.org/abs/1711.07128.

13. Warden, P., Situnayake, D.: *TinyML: Machine Learning with TensorFlow Lite on Arduino and Ultra-Low-Power Microcontrollers*. O'Reilly Media, Incorporated (2020), https://books.google.co.in/books?id=sB3mxQEACAAJ.

14. Banbury, C.R., Reddi, V.J., Lam, M., Fu, W., Fazel, A., Holleman, J., Huang, X., Hurtado, R., Kanter, D., Lokhmotov, A., Patterson, D.A., Pau, D., Seo, J., Sieracki, J., Thakker, U., Verhelst, M., Yadav, P.: Benchmarking TinyML systems: Challenges and direction. CoRR **abs/2003.04821** (2020), https://arxiv.org/abs/2003.04821.

15. Howard, A.G., Zhu, M., Chen, B., Kalenichenko, D., Wang, W., Weyand, T., Andreetto, M., Adam, H.: Mobilenets: Efficient convolutional neural networks for mobile vision applications. CoRR **abs/1704.04861** (2017), http://arxiv.org/abs/1704.04861.

16. ARM: Cortex-m4. Arm Developer (2020), https://developer.arm.com/ip-products/processors/cortex-m/cortex-m4.

17. ARM: Cortex-m7. Arm Developer (2020), https://developer.arm.com/ip-products/processors/cortex-m/cortex-m7.

18. Abadi, M., Agarwal, A., Barham, P., Brevdo, E., Chen, Z., Citro, C., Corrado, G.S., Davis, A., Dean, J., Devin, M., Ghemawat, S., Goodfellow, I., Harp, A., Irving, G., Isard, M., Jia, Y., Jozefowicz, R., Kaiser, L., Kudlur, M., Levenberg, J., Man´e, D., Monga, R., Moore, S., Murray, D., Olah, C., Schuster, M., Shlens, J., Steiner, B., Sutskever, I., Talwar, K., Tucker, P., Vanhoucke, V., Vasudevan, V., Vi´egas, F., Vinyals, O., Warden, P., Wattenberg, M., Wicke, M., Yu, Y., Zheng, X.: TensorFlow: Large-scale machine learning on heterogeneous systems (2015), https://www.tensorflow.org/, software available from tensorflow.org.

19. Chowdhery, A., Warden, P., Shlens, J., Howard, A., Rhodes, R.: Visual wake words dataset. CoRR **abs/1906.05721** (2019), http://arxiv.org/abs/1906.05721.

20. Saied, A.: Fire dataset. Kaggle [Online] (2018), https://www.kaggle.com/phylake1337/fire-dataset.

21. Warden, P.: Speech commands: A dataset for limited-vocabulary speech recognition. CoRR **abs/1804.03209** (2018), http://arxiv.org/abs/1804.03209.

10 Effective Parallel Approach to Enhance the Performance of Cloth Simulation Model for Real-Time Applications

Abhishek S. Rao, Devidas B, and Vasudeva Pai
NITTE (Deemed to be University)

Deepak D. and Nagesh Shenoy
Canara Engineering College

CONTENTS

10.1 INTRODUCTION

In today's era, cloth simulation plays a vital role in fashion industries and e-commerce [1]. When we go shopping for clothes, we want to try them on to ensure that we are purchasing the correct size and that the style is appropriate for our body type. Customers and designers may well see real-life items in motion on their precise virtual models, as well as get fast-fit visualization on the Internet. This allows consumers to see whether the cloth is tight or loose, and how it behaves as they walk around or stand in different poses. Computer-generated garment simulation is the outcome of a wide range of techniques that have grown rapidly over the last decade. Cloth simulation, on the other hand, has progressed to the point where it can be introduced to the clothing business [2]. Virtual clothing prototyping, as well as its applications in fashion industries, are regarded as one of the most important requirements. Virtual clothing design and animation use a variety of approaches, with collision detection and user interface techniques to create clothes. Because of advancements in the development of virtual garment simulation systems, a precise framework that matches the needs of the clothing industry of design and prototyping, emphasizing interactive design, animation, and visualization capabilities, has become a crucial factor to consider. For more than two decades, researchers and practitioners have been developing algorithms and strategies to represent cloth objects realistically on computers. The first cloth simulation model was developed based on the synthesis of cloth objects [3] and its geometrical and physical properties such as elasticity [4], wrinkles [5] effects were proposed for realistic animation outcomes. The cloth surface deforms due to external forces such as gravity or forced twisting, as well as internal forces such as the elastic reaction to tensile load, shearing, and bending. To make the cloth simulation more realistic, it is essential to handle collision between cloth and any geometrical objects such as sphere or cube. In cloth simulation, collision detection has been a substantial problem [6]. Generally, clothing may consist of thousands of vertices and the simulation results will be more realistic; but running on a sequential CPU cannot achieve real-time performance. To address this problem, few researchers have introduced a new data-driven approach by reconstructing the simulation result with a fine-scale animated frame [7]. With the advent of parallel CPU architecture, OpenMP technology has gradually been applied widely in parallel computing. OpenMP helps in achieving various performance factors such as the way the memory is accessed, parallelization overheads, load imbalance overheads, sequential overheads, and synchronization overheads [8].

This paper proposes an effective parallel implementation of cloth simulation using OpenMP. For each cloth vertex, it calculates spring forces, solves mathematical equations, and conducts collision detection in parallel. Experiments demonstrate that our method outperforms other methods, which are sequential by about 2 times and can produce roughly 50–60 animation frames per second, satisfying the real-time performance demand of clothing simulation.

10.2 LITERATURE REVIEW

The realism of the simulated cloth movement and the algorithm speed are two significant areas where clothing simulations might be improved. The researcher has proposed several research problems that are necessary for reaching the desired realism and speed. The author has shed light on the major variables that influence a cloth's characteristics, such as developing non-linear and hysteretic qualities, enhancing the realism of simulated cloth, and increasing algorithm performance while retaining adequate quality. The author also discussed the primary obstacles that collision resolution techniques face, as well as the reasons that make collision handling more difficult [2]. Understanding the properties of fiber mechanics in clothes is critical for virtually simulating clothing. The hierarchical relationships of fabric in the biomechanics function performance of clothes and textile items have been studied by a few researchers [9]. In circumstances where limitations have not been appropriately managed, a heuristic approach was created to model a realistic fabric object [10]. A framework based on the geodesic distance to solve the projected geometric limitations was devised to offer an interactive selection of the distortion influence region and to calculate multiple geometric limitations with interactions [11]. Even though most materials do not stretch when wrapped over the body, numerous approaches based on Constrained Lagrangian Mechanics have been presented to obtain low strain along the warp and weft [12]. Several methods for detecting collisions between a highly deformable item and another object with minimal deformations have been developed. The method works by dividing the space containing the item into cells and assigning an inclusion attribute to each cell [13]. In cloth simulation, the most essential feature that impacts the formation and disappearance of wrinkles and folds is bending behavior. A unique approach known as the dynamic bending model is created from the thin-shell theory to implement the dynamic bending model on spring-mass systems, which is more appropriate to handle huge bending deformation. When compared to previous models, the technique was able to represent the physical properties of woven materials more correctly [14]. A hybrid evolutionary algorithm was utilized to correlate the physical features of fabrics with the elastic modulus of springs that control a model's functionality. The method produced realistic real-life behavior for textile simulations, which could be used in graphical animations and modeling systems to study cloth behavior [15]. The simulation of fabric in virtual worlds based on physical principles is a computationally intensive challenge. For distributed memory parallel systems, a parallel technique was proposed to address this challenge. The authors have covered some parallel strategies for physical modeling and collision handling to lessen the computing load [16].

10.3 METHODOLOGY

In computer graphics, designing a cloth model is a fascinating and challenging subject. The main objective of this work is to design and simulate the cloth by applying the equations of motions and to collide the cloth with a moving ball to achieve a realistic effect.

10.3.1 Design of the Proposed Model

The model comprises of following steps as shown in Figure 10.1.

A cloth can be thought of as a collection of vertices and edges where each vertex will function as a particle and the edges as a spring. The process starts with a two-dimensional triangle mesh as an initial input, which is made up of vertices and edges. The mass characteristics of a lump of cloth surrounding each vertex are concentrated. Each edge represents the material's stiffness and damping characteristics. Internal forces acting on all particles are estimated at each time step of the simulation using the constitutive laws of the material described along the edges. The verlet integration

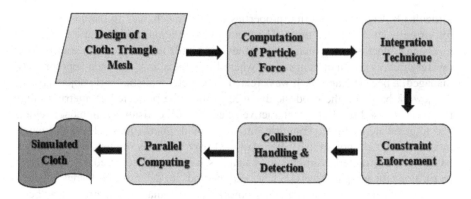

FIGURE 10.1 Proposed design of the model.

method is used to determine the time step integration, and to limit particle motion and fix the distance between particles, constraint enforcement technique is used. The collisions between the particles and the environment are verified, and their positions are adjusted after updating the state of the particles' speed and direction. CPU parallelism will be implemented using the OpenMP parallel programming API for faster simulation.

10.3.2 MASS-SPRING SYSTEM FOR CLOTH

We can simulate the cloth by shifting the mass of the particles and applying force to make them move around in space [9]. To achieve this problem, authors have utilized the technique of a mass-spring-damper system. The spring-mass-damper system is an ideal second-order dynamic system made up of discrete mass nodes spread throughout an item and coupled by a network of springs and dampers. It uses massless elastic springs to describe internal forces between mass locations and computes positions and velocities in discrete time steps [17]. A mass-spring-damper mechanism is used to build the cloth model, with the simulation particles interconnected into a triangle mesh. Each spring-damper joins two particles and generates a force based on their relative locations and speeds. The force of gravity influences each particle, and the foundation of the cloth system will be built using these three simple forces like gravity, spring, and damping. We may then add more complex forces like bending resistance, shearing, and collisions to achieve a more realistic effect. The cloth in the mass-spring model is defined by a mesh of point masses [10]. As illustrated in Figure 10.2, the point masses are interconnected to each other through three different phases of springs such as structural, shear, and bend. In structural springs, a particle indexed by $p(i, j)$ is connected to its neighbor particles indexed by $p(i \pm 1, j)$, $p(i, j \pm 1)$. The shear springs connect the particle to $p(i \pm 1, j \pm 1)$, while the bend springs connect the particle to $p(i \pm 2, j)$, $p(i, j \pm 2)$ and $p(i \pm 2, j \pm 2)$.

As seen in Figure 10.2, each node represents the particle and edges as springs. The diagonal springs are required to prevent the face from collapsing and to prevent the cloth from decomposing into a straight line. The mass of each affected vertex was calculated by taking one-third of the area of each triangle. Gravity vector was

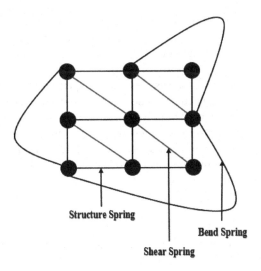

FIGURE 10.2 Mass-spring-damper model.

applied on each particle to describe the gravitational force and computed the force for each spring-damper. As illustrated in Figure 10.3, each mass point in the mass-spring model is operated upon by its weight, damping force, and spring force [18].

Considering mass m, position x, velocity v, the spring force F_s between two particles is derived by Hooke's law [11] as shown in Equation (10.1).

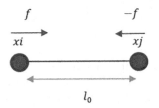

$$F_s = \frac{x_j - x_i}{|x_j - x_i|}\left[k_s(|x_j - x_i| - l_0) + k_d(v_j - v_i) \cdot \frac{x_j - x_i}{|x_j - x_i|}\right]$$ (10.1)

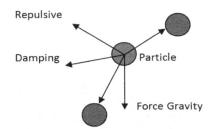

FIGURE 10.3 Particle force analysis.

Here, k_s and k_d are the stretching and damping coefficients, which describe the elastic property of the spring's material, $x_j - x_i$ is the difference between the positions of the two particles, l_0 is the rest length of the spring, and F_s is the restoring force exerted by the spring.

The only effect of gravity on a mass carried on a spring is to define its resting position. Damping is mostly induced by friction with the air, which is always in the opposite direction of motion, and potentially by heat generated by the spring as it flexes. The spring is stretched down to a new equilibrium position by gravity. The upward force generated by this equilibrium stretch balances and negates the gravitational force. When the new equilibrium is disturbed, the mass always pushes back toward it.

The gravitational force F_g is computed as shown in Equation (10.2).

$$F_g = m * g \tag{10.2}$$

Here, m is mass and g is gravity. Linear damping generally takes the form as shown in Equation (10.3).

$$F_{damp} = -b * v \tag{10.3}$$

The (positive) damping constant is b, and the object's velocity is v. As illustrated in Equations (10.4) and (10.5), the repulsive force is produced by elastic force, damping force, and gravity.

$$F_{total} = F_s + F_g + F_{damp} \tag{10.4}$$

$$F_{repulsive} = -k(F_{total} * N)N \tag{10.5}$$

Here, N is normal, and the cloth vertex will be repelled from the geometrical model by scaling k to a value greater than 1. The forces acting on each cloth vertex are estimated using Equations (10.1)–(10.5), without considering the impact of other vertices. To integrate the equation of motion on the cloth model, we must specify a position, velocity, and acceleration for each vertex at each time step t. As an initial stage, authors have applied forces on particles and computed how much it affects the position in a short period. Then, the authors have applied Newton's second law of motion to convert force to acceleration as shown in Equation (10.6).

$$acceleration = \frac{force}{mass} \tag{10.6}$$

10.3.3 INTEGRATION TECHNIQUE

To determine the position of each particle based on the current and previous frames, the authors have employed a numerical integration technique to convert acceleration to position. The type of integration technique that the author used in the research is verlet integration. Verlet integration is an explicit model with the unique virtue of not requiring any prior knowledge of velocity [12]. It calculates the velocity internally by examining the location at both the current and previous time steps as shown in Equation (10.7).

$$x_{t+\Delta t} = 2x_t - x_{t-\Delta t} + \left(\frac{dv}{dt}\right)_t \Delta t^2 \qquad (10.7)$$

Here, x is a position and $\left(\dfrac{dv}{dt}\right)$ is acceleration. Equation (10.7) states that the current location should be offset by the acceleration vector. The result showed us that the current position changes while the old position remains unchanged.

10.3.4 CONSTRAINT ENFORCEMENT TECHNIQUE

High stiffness can lead to instability and may lead to more processing time for the same length of animation. For a stiffness value k_{stiff} of a mass m, a natural period of oscillation T_0 is defined as shown in Equation (10.8).

$$T_0 \approx \pi \sqrt{\frac{m}{k_{stiff}}} \qquad (10.8)$$

The timestep Δt must be less than T_0. To avoid the superelastic effect without excessively decreasing Δt, authors have employed a constraint enforcement mechanism to limit particle motion and fix the distance between particles after one integration step such that particles stay in a grid [13]. Particle shift during verlet integration causes particles to be too far apart or too close together. As a result, the authors use a constraint satisfaction technique to change the position of the two particles so that their distance returns to its original length. At the initial stage, the authors calculated the deformation rate of each spring and verified whether the deformation rate is greater than the critical deformation rate. If the deformation rate is greater, then the authors have applied a dynamic inverse procedure to limit the deformation to critical deformation rate so that the springs should not exceed more in length as shown in Figure 10.4.

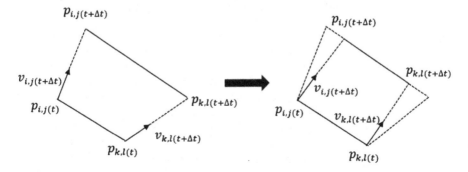

FIGURE 10.4   Adjustment of a spring connecting two loose masses.

The shear/bend constraints helped in retaining the cloth's shape, while the stretch constraints made the particles intact to each other. The stretch constraint in the cloth simulation model is proportional to the size of the spring force. Increased spring stiffness required smaller time steps during numerical integration; however, altering the vertex positions of the cloth for each iteration reduces the overall performance of the simulation.

10.3.5 COLLISION DETECTION AND HANDLING

To make the cloth behave dynamically, authors have applied wind forces, gravity, and collision with a moving ball. As the cloth is viewed as a collection of three-particle triangles, wind can be applied to one triangle at a time by applying forces to each of the three particles. Although the wind can originate from any direction, the force was only added in the direction of the triangle's normal vector. The amount of force acting on the triangle was made proportionate to the triangle's angle with the wind. The gravitational force was then applied to all particles by adding a velocity vector.

Collisions are very hard to deal with when it comes to cloth [6]. To collide an object with a cloth, we need to determine the $x(t+dt)$ starting at a time 't' and cloth position $x(t)$. Considering the beginning and end positions of each face on the mesh, authors have computed the time when these surfaces came in touch between time t and $t+\Delta t$. If the face was previously in contact with one and is no longer in contact, then we need to disconnect the surfaces. Static and kinetic friction forces were calculated that contacted the surface and continued the simulation from this new time step. The ball is defined by a center and a radius and visualized using the OpenGL graphics library. If the particle is inside the radius of the ball, the collision is detected [19] and handled by shifting the particle out from the ball along the vector from center to particle until the distance is equal to the ball's radius, as shown in Figure 10.5.

Its surface is defined by the following Equation (10.9).

$$|p - C| = r \qquad (10.9)$$

Here, p is any point on the surface of the sphere. For points inside the sphere, we have $|p - C| < r$ and for points outside the sphere, we have $|p - C| > r$.

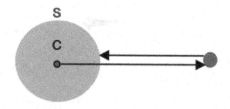

FIGURE 10.5   Sphere-cloth collision.

10.3.6 OpenMP Parallel Programming Model

OpenMP is one of the standards for parallelizing programs in a shared memory environment. The compiler directives are the primary tools for parallelizing code in OpenMP and are used in the program to indicate which block of the code must be parallelized. The fork-join concept is used in OpenMP to parallelize an application that has interleaved serial and parallel sections. Starting with a serial program, the OpenMP will allow us to selectively parallelize the most time-consuming areas of the program until we obtain adequate performance. An OpenMP program starts with a single execution thread, known as the initial thread and when a thread encounters a parallel construct, it forms a new team of threads, consisting of itself and more additional threads. The block of code is shared by all the threads within the parallel construct, and only the master thread remains to run user code after the parallel construct has stopped [20].

OpenMP has several environment variables that can be used to alter how the program behaves. Since each cloth vertex has little association with others in cloth simulation based on the mass-spring paradigm, the OpenMP threads were employed to parallelize the calculation of forces, collision handling, and position updates for each mass point. The forces acting on each cloth vertex can be estimated separately using Equations (10.1)–(10.5), without taking other vertices into account. As a result, authors have employed the OpenMP parallel computing approach to calculate the spring force, gravity, damping force, and repulsive force parallelly thereby boosting the model's performance.

10.3.7 Framework for Physical Simulation

The OpenGL framework is constructed on top of the windowing framework, and the physical simulation framework is developed on top of all that. To make the application run in real-time, time-based rendering is used. The time it takes to calculate each frame is referred to as this delay. Time-based rendering also assures that the simulation's rate of progression is unaffected by the system's speed. This contrasts with frame-based rendering, which involves drawing and displaying new frames as quickly as the system can generate them. For modeling things in three-dimensional space, the physical simulation framework employs a particle-based approach. Position, velocity, mass, and the sum of the forces acting on them are all attributes of particles. The sum of forces is set to zero at the start of each time step and forces are applied to particles. Once the forces for each particle have been determined, the force sum, together with the mass of the particle, must be used to compute the new position and velocity.

10.4 RESULTS

The cloth simulation model was written in C++ using the OpenGL graphics library and ran on a desktop computer with an Intel® Core™ i7-2600 processor running at 3.40 GHz and 4 GB of RAM. To compare the performance of the simulation for each mesh size, the cloth is simulated using sequential (CPU) and

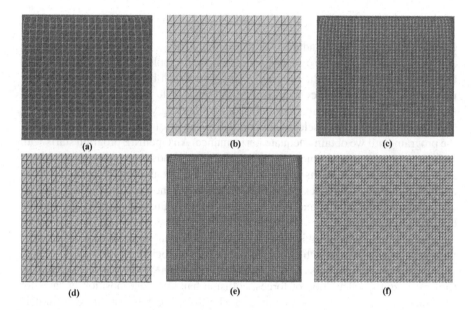

FIGURE 10.6   The cloth mesh: (a) 20×20 (b) 30×30 (c) 40×40 (d) 50×50 (e) 60×60 (f) 70×70.

parallel (OpenMP) approach. The six mesh sizes that were observed are depicted in Figure 10.6. The lower bound mesh was chosen as 20×20, 30×30, 40×40, and the higher bound mesh were chosen as 50×50, 60×60, 70×70. The sequential and parallel CPU frame rates for different modes of cloth simulations per second are shown in Figure 10.7.

It is evident from Figure 10.7 that in the high parallel phase, OpenMP can achieve a 2–3 times speedup ratio when compared to CPU alone. Figure 10.8 shows the sequential versus OpenMP simulation speedup ratio (%).

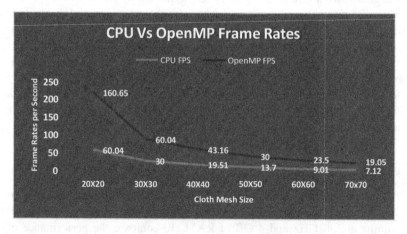

FIGURE 10.7   Frame rate ratio for 30×30, 50×50, and 70×70 meshes.

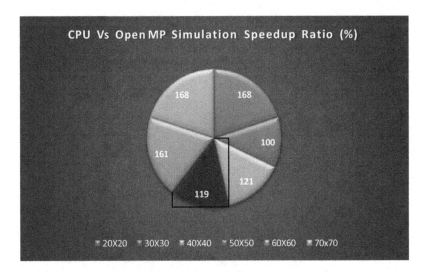

FIGURE 10.8   CPU vs OpenMP simulation speedup ratio.

We can see from the findings that OpenMP can significantly improve performance in high parallel modules. Figure 10.9 shows the OpenGL view of the cloth sheet's springs, polygons, folded cloth, and sphere-cloth collision detection.

10.5   DISCUSSION

Instead of applying velocity to the particles, position-based dynamics can be created by altering positions directly. If penetration of any object is detected, we can move the objects to enforce non-interpenetration and update to appropriate velocities instead of delivering a force or impulse [21]. Shaders can also do particle collision and updating. Collisions are identified and treated by pushing the particle out of the collided volume. Spring restrictions and forces are evaluated for each cloth particle, and collisions are detected and handled for every interactable scene geometry [22].

10.6   CONCLUSION

This research shows how a user may use a Sequential – Parallel CPU computing architecture for clothing simulation to animate a virtual cloth in real-time. The study also discusses how using the OpenMP concept with the simplified mass-spring model improved computational efficiency. To achieve a realistic simulation effect, the cloth collided with a ball, and its performance is calculated based on frame rates for each simulation mode. When compared to a sequential CPU, this framework uses OpenMP to handle high-level parallel modules like force computation, constraint satisfaction, and numerical integration methods, resulting in speedup ratios of 2–3 times. This approach can be further enhanced by adopting GPU programming utilizing CUDA, OpenCL, and OpenACC to meet the growing demand for fast cloth simulation applications.

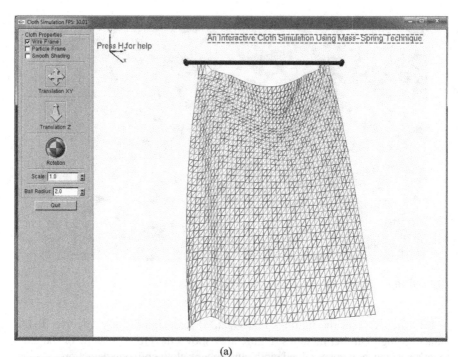

(a)

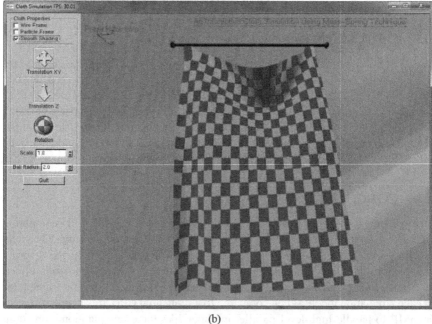

(b)

FIGURE 10.9 Cloth simulation: (a) OpenGL view of the cloth sheet's springs. (b) OpenGL view of the cloth sheet's polygons. (c) Folded cloth. (d) Sphere-cloth collision detection.

(*Continued*)

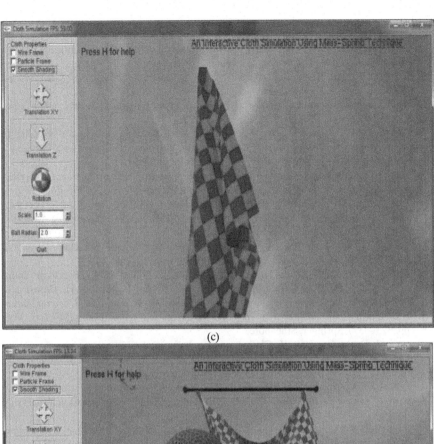

(c)

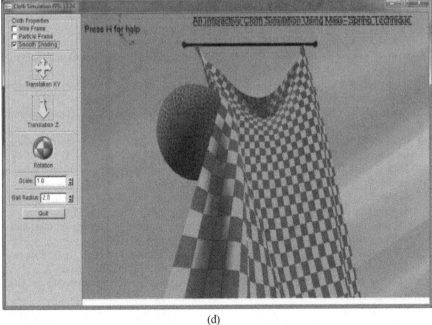

(d)

FIGURE 10.9 (*Continued*)   Cloth simulation: (a) OpenGL view of the cloth sheet's springs. (b) OpenGL view of the cloth sheet's polygons. (c) Folded cloth. (d) Sphere-cloth collision detection.

REFERENCES

1. Zhang X, Wong LY. Virtual fitting: real-time garment simulation for online shopping. ACM SIGGRAPH 2014 Posters 2014 Jul 27 (pp. 1–1).
2. Choi KJ, Ko HS. Research problems in clothing simulation. *Computer-Aided Design.* 2005; 37(6): 585–92.
3. Weil J. The synthesis of cloth objects. *ACM Siggraph Computer Graphics.* 1986; 20(4): 49–54.
4. Terzopoulos D, Platt J, Barr A, Fleischer K. Elastically deformable models. In *Proceedings of the 14th Annual Conference on Computer Graphics and Interactive Techniques,* ACM, New York, 1987 Aug 1 (pp. 205–214).
5. Kunii TL, Gotoda H. Singularity theoretical modeling and animation of garment wrinkle formation processes. *The Visual Computer.* 1990; 6(6): 326–36.
6. ElBadrawy AA, Hemayed EE, Fayek MB. Rapid collision detection for deformable objects using inclusion fields applied to cloth simulation. *Journal of Advanced Research.* 2012; 3(3): 245–52.
7. Feng WW, Yu Y, Kim BU. A deformation transformer for real-time cloth animation. *ACM Transactions on Graphics (TOG).* 2010; 29(4): 1–9.
8. Chandrashekhar BN, Sanjay HA. Performance study of OpenMP and hybrid programming models on CPU–GPU cluster. In: Shetty, N., Patnaik, L., Nagaraj, H., Hamsavath, P., Nalini, N. (eds), *Emerging Research in Computing, Information, Communication and Applications,* Advances in Intelligent Systems and Computing, 2019 (vol. 906, pp. 323–337). Springer, Singapore.
9. Long J, Burns K, Yang JJ. Cloth modeling and simulation: a literature survey. In: Duffy, G., Vincent (eds), *International Conference on Digital Human Modeling,* 2011 (pp. 312–320). Springer, Berlin, Heidelberg.
10. Provot X. Deformation constraints in a mass-spring model to describe rigid cloth behaviour. In *Graphics Interface,* 1995 (pp. 147–147). Canadian Information Processing Society, Québec.
11. Giuliodori MJ, Lujan HL, Briggs WS, Palani G, DiCarlo SE. Hooke's law: applications of a recurring principle. *Advances in Physiology Education.* 2009; 33(4): 293–6.
12. Cetinaslan O. Localized constraint based deformation framework for triangle meshes. *Entertainment Computing.* 2018; 26: 78–87.
13. Goldenthal R, Harmon D, Fattal R, Bercovier M, Grinspun E. Efficient simulation of inextensible cloth. ACM SIGGRAPH 2007 papers 2007 Jul 29 (pp. 49-es).
14. Zhou C, Jin X, Wang CC, Feng J. Plausible cloth animation using dynamic bending model. *Progress in Natural Science.* 2008; 18(7): 879–85.
15. Mongus D, Repnik B, Mernik M, Žalik B. A hybrid evolutionary algorithm for tuning a cloth-simulation model. *Applied Soft Computing.* 2012; 12(1): 266–73.
16. Thomaszewski B, Blochinger W. Physically based simulation of cloth on distributed memory architectures. *Parallel Computing.* 2007; 33(6): 377–90.
17. Eshkabilov SL. Spring-mass-damper systems. In: Eshkabilov, S. L. (ed), *Practical MATLAB Modeling with Simulink,* 2020 (pp. 295–344). Apress, Berkeley, CA.
18. Neelima B, Rao AS. Effective algorithm for improving the performance of cloth simulation. *Proceeding of NACCET'12,* Bengaluru, 2012; pp. 170–175.
19. Sushma MD, Rao AS. Collision detection for volumetric objects using bounding volume hierarchy method. *International Journal of Computer & Mathematical Science,* 2014; 3(2): 42–51.
20. Chapman B, LaGrone J. OpenMP. In: Pankratius, V., Adl-Tabatabai, A-R., Tichy, W. (eds), *Fundamentals of Multicore Software Development,* 2012 (pp. 113–140). CRC Press, Boca Raton, FL.

21. Müller M, Heidelberger B, Hennix M, Ratcliff J. Position based dynamics. *Journal of Visual Communication and Image Representation*. 2007; 18(2): 109–18.
22. Zeller C. Cloth simulation on the GPU. ACM SIGGRAPH 2005 Sketches 2005 Jul 31 (pp. 39-es).

11 Pruning in Recommendation System Using Learned Bloom Filter

Harsha Verma, Aditya Manchanda,
and Amrit Pal Singh
Dr. B.R. Ambedkar National Institute of Technology

CONTENTS

11.1 INTRODUCTION

One of the primary reasons for the need for a recommender system (RS) in modern society is that people have so many choices due to the widespread use of the Internet. Earlier, people used to shop in physical shops where the range of products was minimal. The number of movies that can be stored in a Blockbuster store, for example, is determined by the store's size. An RS is a standard tool that provides recommendations for objects, goods, content, or services that users may enjoy. RS compiles a ranked list of things that a consumer may be interested in. Movies, books, groups, news, and articles all have their RS. They are intelligent applications that aid users in making decisions when they are faced with a potentially overwhelming number of alternative products or services [1]. We have an object and a user in a recommendation framework. To recommend a product to a customer, any RS requires some parameters such as rating or item attributes [2]. By giving specific recommendations to users, the recommender framework assists them in locating relevant and related products. These systems scour the Internet for specific recommendations for users based on their specifically stated interests or objective habits [3]. RSs are designed to provide customers with a list of products that a service recommends. A video streaming service, for example, will usually use an RS to provide each of its customers with a customized list of movies or shows. Rating prediction in recommendation aims to predict the missing ratings for all remaining user-item pairs given an incomplete dataset of user-item interactions that take the form of numerical ratings (e.g., on a scale from 1 to 5) [4]. An RS, in general, is any software system that actively recommends an item for purchase, subscription, or investment. An advertisement can be viewed as a suggestion in this broad context. However, we concentrate on a more restricted concept of "personalized" RS that is focused on user-specific data [5]. Knowledge discovery techniques are applied to the problem of making customized recommendations for information, goods, or services during a live interaction by RS [6]. These systems, especially those based on K-nearest neighbor collaborative filtering (CF), are gaining traction on the Internet [6]. The amount of work in conventional CF systems rises concerning the number of participants in the system. New RS technologies are needed to deliver high-quality recommendations quickly, including for very large-scale problems [6]. As a result, an RS must employ suitable filtering techniques for certain objects to be ranked, either implicitly or explicitly, and thus recommended. The two most common types of RSs are content-based RSs and CF, which use vastly different algorithms to produce recommendations [7]. The content-based recommender framework compares the user's previously purchased or searched items and suggests related items [2]. A content-based recommender framework recommends products that have features in common with items the user has previously enjoyed. A traditional content-based recommender starts by creating a user profile based on user input and item ratings [3]. A pure content-based RS only considers the user's preferences; however, a system must also consider the preferences of the user's friends. This can improve prediction accuracy and provide user satisfaction. Things that are not searched by the user but may be useful to the user are not recommended by the content-based RS. This is referred to as a serendipity problem. For prediction, many researchers have used a variety of algorithms. For prediction, many researchers have used a variety of algorithms. For example, the K-nearest neighbor algorithm compares the new item to be recommended against all previously stored items and finds the K closest neighbors or items that are identical to the new item. Decision Tree, Naive Bayes Classifier, and ECSN are some of the algorithms used in content-based RSs [2,7]. Different profile-item matching strategies are used in the content-based approach to compare features of new products with user profiles and determine whether a specific item is interesting to the user. Utilities are allocated to objects based on utilities previously assigned to observed items, and user profiles are either implicitly or specifically modified. Items are automatically ranked by relevance and irrelevance by comparing item views with user interests and interests using keyword match, search focus for neighbors closest, cosine similarity, and typical ratings. CF recommends items by matching users with other users with similar interests. Collect user feedback as user-submitted ratings for a particular item and compare rating behavior among users to find groups of users with similar attitudes. A user profile is a user setting that the user has explicitly or implicitly specified [3]. Active user prediction is obtained by taking the weighted average of all similar user ratings. The accuracy of the recommendation system depends on how good it is to select other users who are similar to the user who wants to receive recommendations. CF algorithms must handle many users [2]. The collaborative algorithm is one of the most successful and widely used social recommendation algorithms. It was first proposed in 1992 by Goldberg, Nichols, Okey, and Terry. Using a collaborative method, a system called Tapestry was created to filter emails. It can only cover a small number of users and requires user attributes, which is a new recommendation method. Unlike content-based recommendation algorithms, CF does not require identification of the item's content, it should capture users who have rated the item, and ultimately users should receive recommendation results. Collect reviews from many users on social networks and filter out chaotic and irrelevant information. It has been applied to e-commerce and personalized online communities [8]. Researchers have developed many CF algorithms. These algorithms can be divided into two categories: memory-based and model-based algorithms [6]. Memory-based CF is a sort of early-generation CF that uses heuristic algorithms to generate similarity values between persons or objects. It may be classified into two types: CF based on users and CF based on items. Memory-based algorithms make predictions based on the entire user-item database. These systems use statistical methods to identify a group of users, referred to as neighbors, who have previously agreed with the target user. These systems combine the preferences of neighbors to generate a prediction or top-N recommendation for the active user once a neighborhood of users has been created. The techniques are more common and commonly used in practice and are also known as nearest-neighbor or user-based CF. Model-based CF, as opposed to heuristic techniques, creates a model to anticipate a user's rating on things using machine learning or data mining methods. The recommendation is provided by model-based CF algorithms, which begin by creating a model of user ratings. The CF method is envisioned as computing the expected value of a user prediction, given his/her ratings on other objects, by algorithms in this category [6]. Apart from this, there are some other RSs that are: Hybrid RSs provide recommendations by combining various methods of recommendation. Rather than using a single technique, it can provide better results [2,9]. In a demographic-based RS, the user's characteristics, such as gender, age, and education are considered. There is no requirement for the product ratings or feedback given by users. As a result, first-time users could receive recommendations even if they have not given any ratings to any of the products on the website [10]. In a keyword-based recommendation system, user preferences are indicated by keywords. The keyword recommendation engine uses user-provided textual overviews to generate recommendations and uses the MapReduce framework to provide more scalable execution in a big data environment with two types of users: previous users and active users. Keywords are based on previous user ratings. The extracted keywords are stored in the database. There are several ways to extract keywords. Active users enter keywords, and the meaning of keywords can be expressed from 1 to 5. Here, we identify comments from previous users that are similar to comments from active users.

Therefore, we need to calculate the similarity between keywords in previous user ratings and keywords reported by active users [2].

The proposed system pruning in RS using learned bloom filter (BF) provides a solution to the irrelevant, redundant, scalability, and space issues. These issues are very common in normal RS. Irrelevant results mean giving recommendation, which is not relevant or useful for the user, redundant means to provide same recommendation again and again to a user, scalability issues come when the algorithm need to search a large amount of data. Nowadays, RSs are almost being used in many areas so the problem of handling a huge amount of data is challenging and the traditional system fails in that, so the probabilistic data structure (PDS) is used for solving this issue.

The rest of this chapter is arranged as follows: Section 11.2 explains related work in RS using different recommendation algorithms and RS using PDS to provide a recommendation. Section 11.3 describes the background and context this section explains about PDS and BF. Section 11.4 introduces challenges in RS. Section 11.5 describes the proposed work. Section 11.6 describes observation and analysis. Section 11.7 shows the graph of the result obtained. Section 11.8 conclusion and future scope.

11.2   RELATED WORK

In this section, we briefly review how RS is being used in diverse fields, proposed in the past few years. The researchers working in diverse areas to explain how the RS is working, how important the RS is in today's time for everyone like an e-commerce site and an individual, and how the algorithm of RS is used to provide the recommendations and discussing the problems that they are facing and some of the solutions have also been given by them. So, to carry out the research work, we have selected some of the research paper related to RS its algorithm and some ML approach has been used for better understanding of the work.

The huge development in the measure of accessible data and the number of guests to Web destinations in late years represents some vital difficulties for recommender frameworks. These are: creating excellent proposals, performing numerous suggestions each second for millions of clients and things, and accomplishing high inclusion even with information sparsity. In conventional collective filtering frameworks, the measure of work increments with the number of members in the framework. New recommender framework innovations are required that can rapidly create excellent proposals, in any event, for exceptionally enormous scope issues. Sarwar et al. [6] examine various item-based recommendation generation algorithms in this paper. They investigate various methods for computing item-item similarities extracting suggestions from them. Their findings indicate that item-based algorithms outperform user-based algorithms while still offering higher efficiency than the best possible user-based method.

Collective filtering is a quickly propelling examination region. Consistently, a few new methods are proposed but then it isn't clear which of the procedures work best and under what conditions. Lee et al. [5], in this paper, lead an investigation contrasting a few cooperative sifting strategies – both work of art and ongoing cutting edge – in an assortment of test settings. In particular, the report ends controlling for the number of things, number of clients, sparsity level, execution models, and computational intricacy. Their results demonstrated which technologies perform well and under what circumstances, and they benefit both the manufacturing and research communities by identifying which CF algorithms work well and under what circumstances.

Recommender frameworks can offer significant types of assistance in a computerized library climate, as exhibited by its business achievement in the book, film, and music enterprises. Perhaps the most usually utilized and fruitful suggestion calculation is synergistic sifting, which investigates the connections inside client thing communications to deduce client interests and inclinations. In any case, the suggested nature of collective sifting approaches is incredibly restricted by the information sparsity issue. To lighten this issue, Huang et al. [11] have recently proposed chart-based calculations to investigate transitive client thing affiliations. They expand the concept of examining user–item interactions as graphs in this paper, and they use connection the systematic method proposed in recent network modeling research to make CF suggestions. For making suggestions, they've implemented a variety of linkage interventions. Both path-based and neighbor-based approaches perform better than traditional user-based and item-based algorithms, according to the findings.

Dou et al. [8] with other researchers mentioned some hybrid methods that could be used to boost the algorithm's efficiency, but there is still scope to introduce more approaches or update the existing algorithms. The paper proposed by them examines the CF algorithm's three key problems and the relevant research that has been done to address them. Their outcome provides multiple solutions that can be used to solve the problem of cold start and to minimize the value of finest neighborhood computing.

A recommender framework is one of the significant strategies that handle the data over-burden issue of Information Retrieval by proposing clients with suitable and important things. Today, a few recommender frameworks have been created for various areas; notwithstanding, these are not exact enough to satisfy the data needs of clients. Along these lines, it is important to fabricate top-notch recommender frameworks. In planning such recommenders, fashioners face a few issues and difficulties that need appropriate consideration. Khushro et al., [3] in this paper, examine the latest events, problems, obstacles, and research directions in developing high-quality RS. Explored what can be done to alleviate these problems, as well as what needs to change in the form of various research opportunities and recommendations that can be followed in dealing with issues such as latency, sparsity, background knowledge, grey sheep, and the cold start problem.

Although recommender systems in a broad sense remain an active field of research, research on CF algorithms based exclusively on scores is decreasing. Kluver et al., [12] in this chapter, discussed the key ideas, algorithms, and assessment methods in ratings-based CF. While there are numerous proposal calculations, the ones they cover to fill in as the reason for a lot of at various times calculation advancement. In the wake of introducing these calculations, they present instances of two later bearings in suggestion calculations: figuring out how to rank and group proposal calculations. They conclude this paper by explaining how CF algorithms can be tested, as well as several tools and datasets that can be used to help future research.

The method of shared separating is particularly effective in producing customized suggestions. Over time of exploration has brought about various calculations, albeit no correlation of the unique techniques has been made. Indeed, a generally acknowledged method of assessing a collective sifting calculation doesn't exist yet. Cacheda et al., [13] in this work, examine the characteristics of various strategies reported in the literature and contrast them, emphasizing their major strengths and weaknesses. Several trials were carried out with the most widely used measurements and techniques. In addition, two new measures for assessing the accuracy of good products have been proposed. In the cases broke down, its outcomes are at any rate comparable to those of the best methodologies examined. Under sparsity conditions, there is over a 20% improvement in precision over the customary client-based calculations, while keeping up more than 90% inclusion.

CF can be isolated into two primary branches: memory-based and model-based. Most of the present explores works on the exactness of memory-based calculations exclusively by further developing the likeness measures. In any case, not many scientists zeroed in on the expectation score models, which we accept are a higher priority than the closeness measures. The most notable calculation to demonstrate memory-based is the network factorization. Contrasted with memory-based calculations, the lattice factorization calculation, by and large, has higher exactness. Notwithstanding, the framework factorization might fall into neighborhood ideal in the learning cycle, which prompts lacking learning. CF approaches are generally intended to give items to expected clients. Accordingly, the precision of the strategies is vital. Zhang et al. [9] propose different answers to make a quality suggestion. The strategies they referenced in this paper are identified with numerous communitarians separating procedures that incorporate the network factorization procedures and the neighbor-based techniques. The presented hybrid model has been shown to improve suggestion quality. They can observe that their hybrid technique outperforms DBRISMF and DVIIin terms of accuracy. The results of this study using the MovieLens dataset proved the efficacy of our approach.

Today, recommenders are generally utilized for different purposes, particularly managing Web-based business and data separating apparatuses. Content put together, recommenders depend on concerning the idea of closeness between the purchased, looked, visited thing, and every one of the things put away in an archive. It is a typical conviction that the client is keen on what is like what she has as of now purchased, looked, and visited. All of us accept that there are a few settings where this supposition isn't right: it is the situation of gaining unsearched yet valuable things or snippets of data. This is called luck. Iaquinta et al. [14] present the plan and execution of a crossover recommended framework that joins a substance-based methodology and fortunate heuristics to alleviate the overspecialization issue with amazing ideas The findings show a pattern: the higher the threshold of unpredictability, the higher the number of excellent evaluations. This might be understood as follows: the randomness of the fortuitous item selection aids in the improvement of ratings.

The majority of the current recommender frameworks for the travel industry apply information-based and content-based methodologies, which need adequate verifiable rating data or additional information and experience the ill effects of the cold start issue. Wang et al., [10] in this research, used a demographic recommender system to make attraction recommendations. This algorithm classifies travelers based on their demographic data and then offers recommendations based on those categories. Its benefit is that it does not require a history of ratings or further expertise, so a new traveler can get a recommendation. According to the findings of the experiments, three machine learning approaches based on demographic data outperformed the baseline technique, particularly the SVM approach. These initial findings show that demographic data on its own is insufficient for accurate rating prediction, while further research is needed to corroborate our findings, such as managing imbalanced data and evaluating different types of attractions.

Su and Khoshgoftaar, [15] in this paper, initially present CF assignments and their principal challenges, like information sparsity, versatility, synonymy, dark sheep, peddling assaults, and security insurance, and their potential arrangements. They then, at that point, present three primary classifications of CF procedures: memory-based, model-based, and mixture CF calculations with models for agent calculations of every class, and examination of their prescient exhibition and their capacity to address the difficulties. From fundamental procedures to best in class, they endeavor to introduce an extensive review of CF strategies, which can be filled in as a guide for exploration and practice around here. Through this survey, they have given that, even though there is no fix all arrangement accessible yet, individuals are working out answers for every one of the issues.

Suggestion calculations are most popular for their utilization on Internet business Web locales, where they utilize input about a client's advantages to create a rundown of suggested things. Numerous applications utilize just the things that clients buy and unequivocally rate to address their inclinations, yet they can likewise utilize different characteristics, including things saw, segment information, subject interests, and most loved craftsmen. Linden et al. [16] suggest that conventional CF, cluster models, and search-based algorithms are three typical approaches to tackling their recommendation problem. They contrast these techniques to their technique, which they name item-to-item CF, in this publication. Unlike conventional CF, their method's online computing scales in proportion to the number of customers and products in the catalog. The algorithm's suggestion value is outstanding, because it recommends strongly correlated similar things. Unlike typical CF, the algorithm can deliver high-quality suggestions based on as few as 2 or 3 items, unlike standard CF.

Recommender frameworks further develop admittance to pertinent items also, data by making customized ideas based on past instances of a client's preferences. Most existing recommender frameworks utilize shared sifting techniques that base suggestions on other clients' inclinations. Things that have not been evaluated by an adequate number of clients can't be successfully suggested. On the other hand, content-based strategies use data about a thing itself to make ideas. This methodology enjoys the benefit of having the option to prescribe already unrated things to clients with exceptional interests and to give clarifications to its proposals. Mooney et al. [17] have fostered a book suggesting a framework that uses semi-organized data about things accumulated from the Web utilizing straightforward data extraction methods. The results of the first experiments show that this method can give reasonably accurate recommendations. Generally, the forecasts are sensibly exact even given moderately few preparing sets (25 models). Moderate relationships (above 0.3) are created after around 20 models and solid connections (above 0.6) after around 60 models. While the parallel model beat both the 10-evaluations model and the weighted paired model for a twofold forecast on informational collection 1, the distinction between any of the models isn't measurably huge.

Among different conceptualizations, shrewd urban areas have been characterized as practical metropolitan regions explained by the utilization of information and communication technologies (ICT) and current frameworks to confront city issues in effective and economical manners. Inside ICT, recommender frameworks are solid apparatuses that channel important data, redesigning the relations between partners in the commonwealth and common society, also, aiding dynamic assignments through mechanical stages. There are logical articles covering suggestion approaches in brilliant city applications, and there are proposal arrangements carried out in certifiable savvy city drives. Quijano-Sanchez et al., [18] in this paper, have overviewed the examination writing on recommender frameworks for shrewd urban areas, introducing a portrayal, order, and near examination of distributed papers. Because of their overview, they don't just distinguish and examine primary exploration drifts yet additionally show current freedoms and difficulties where customized proposals could be misused as answers for residents, firms, and public organizations.

In recent years, a lot of examination exertion has been dedicated to creating calculations that produce suggestions. The subsequent exploration progress has set up the significance of the client thing network, which encodes the individual inclinations of clients for things in an assortment, for recommender frameworks. The U-I lattice gives the premise to synergistic sifting strategies, the prevailing structure for recommender frameworks. As of now, new proposal situations are arising that offer promising new data that goes past the U-I lattice. Shi et al., [19] in this study, sum up and analyze proposal situations including data sources and the CF calculations that have been as of late created to address them. They give a far-reaching prologue to an enormous assortment of examination, over 200 key references, determined to help the further advancement of recommender frameworks abusing data past the U-I lattice. Based on this material, they recognize and talk about what they see as the focal difficulties lying ahead for recommender framework innovation, both as far as expansions of existing strategies just as of the mix of methods and advances drawn from other examination regions.

The essential assignment of the memory-based CF proposal framework is to choose a gathering of closest (comparative) client neighbors for a functioning client. Conventional memory-based CF plots will in general just spotlight on working on however much as could reasonably be expected exactness by suggesting recognizable things. However, this might decrease the number of things that could be suggested and subsequently debilitates the shots at suggesting novel things. Karabadji et al. [20] center primarily around the development of the enormous inquiry space of clients' profiles and utilize a transformative multi-target streamlining-based proposal framework to pull up a gathering of profiles that boosts the two similitudes with the dynamic client and variety between its individuals. The trial results on the MovieLens benchmark and a certifiable protection dataset show the effectiveness of our methodology as far as exactness and variety looked at to best-in-class contenders.

Notwithstanding the inadequate client thing (U-I) network, an expanding number of current recommender frameworks look to further develop execution by taking advantage of extra heterogeneous information sources. Such rich side sources can give extremely helpful data about clients' very own practices and things' properties and, accordingly, can essentially profit recommender frameworks. Most existing work can just fuse a solitary side wellspring of clients or things. Chen and Li [21] propose a novel calculation, MSUI, that can at the same time think about different side wellsprings of clients and things in a brought together system. To accomplish this, they use a joint framework factorization calculation all while catching the client/thing dormant elements from the U-I lattice and different side sources. For each dataset, they haphazardly select 40%, 60%, and 80% of the evaluations as the preparation sets, and the rest as the testing sets. The arbitrary determination is done multiple times autonomously, and the normal RMSE is accounted for. In the trials, they observationally set $\alpha_i = 0.01$ for all perspectives on clients and $\beta_j = 0.01$ for all perspectives on things (e.g., Yelp dataset).

11.3 BACKGROUND AND CONTEXT

In the modern age of technology, statistical analysis and data mining of gigabyte-sized databases of several terabytes of data have become routine. Since the data is too large to accommodate in storage, deterministic data structures are not possible when the datasets in which a program deals become quite large. Streaming applications, which usually require data to be processed in a single pass and incremental updates, make it much more complicated. Traditional methods can produce reliable results, but they take time and space. Probabilistic alternatives to deterministic data structures are better in terms of complexity and continuous factors involved in an actual runtime.

11.3.1 PDS

Late examination bearings in the space of big information preparing, investigation, and representation show the significance of PDS.

A deterministic data structure that is in use, like an array, list, hash table, etc. These in-memory information structures are the most commonplace information structures on which different tasks, for example, addition, discovery, and erase could be performed with explicit key qualities. Because of the activity what we get is the deterministic (accurate) result. In any case, this isn't on account of a PDS, Here, the consequence of activity could be probabilistic (may not offer you a positive response, consistently brings about inexact), and thus named as a probabilistic information structure. PDS works with an enormous informational index, where we need to play out certain activities, for example, discovering some exceptional things in each informational collection or it very well may be finding the most successive thing or if a few things exist or not. To do such an activity, probabilistic information structure utilizes increasingly more hash capacities to randomize and speak to a bunch of information. The utilization of deterministic information constructions to examine in-stream information frequently incorporate a lot of computational, space, and time intricacy. Probabilistic options PDS to deterministic information, structures are better as far as effortlessness and consistent components engaged with a real runtime. They are reasonable for enormous information handling, estimated forecasts, quick recovery, and putting away unstructured information, in this manner assuming a significant part in big information preparing [22].

11.3.2   BF

A BF is a bit-architecture that represents n-elements of the same set S in a smaller space of m-bits (See Figure 11.1). The m-bits at first are set to 0, reflecting the missing elements in the filter. The insertion of elements into the bit structure is then efficiently distributed by k-independent hash functions. This affects the status of the structure's k pieces. Particularly, to place an element, the structure's k-positions are hashed to set to 1. The BF structure enables quick membership queries. The requested element is hashed with the k hash functions, and the resulting indexes are checked in the bit structure. The BF ensures that the element that has not been inserted in any of these indexes is set to 0. When every index is put to 1, it is reasonable to conclude that the element has been inserted with a false-positive probability. A false positive occurs when a membership query returns that the element belongs to the set, but it does not. Nonetheless, this false-positive ratio can be estimated, and close-choice boundaries can be calculated [23].

In this research work for giving recommendations more effectively and to handle huge data, BF, which is a PDS, is used as the deterministic data structure, which is not that effective in handling large data. After reviewing different research areas where RS is used using different methods and every researcher have proposed different-different model to make the system more efficient, still the system has multiple challenges that need to be overcome in future.

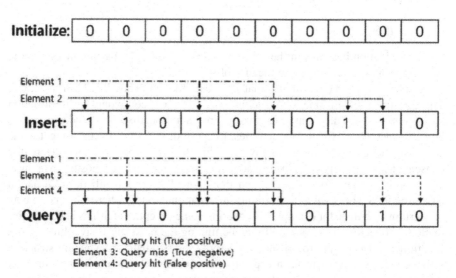

FIGURE 11.1   Operations on BF.

11.4 CHALLENGES IN RS

- **Cold Start Problem:** This issue occurs when fresh patrons join the system or fresh objects are added to the catalog. In such cases, neither the fresh patrons' tastes nor the fresh objects can be rated or bought by the patrons, resulting in less precise recommendations [6].
- **Scalability**: CF algorithms must search through millions of patrons. The website contains a wealth of facts. So, this is the scalability issue [2].
- **Grey Sheep**: Many times, it is easy to gather patrons who have similar likes and dislikes. Some users may not share enough likes or dislikes. These patrons are referred to as grey sheep patrons. As a result, precise recommendations may be impossible to make in such a case [2].
- **Sparsity**: The availability of large amounts of data about items in the catalog, combined with users' reluctance to rate items, results in a dispersed profile matrix, resulting in less precise recommendations. The sparse rating in CF systems makes it difficult to make accurate predictions about items. CF recommends objects based on their closest neighbors, and with fewer ratings, calculating neighbors becomes computationally difficult [6].
- **Overspecialization**: Recommendations are all the same. Only items similar to those previously rated or purchased will be recommended to the user. This is also known as the serendipity dilemma [14].
- These systems are difficult to implement as well as costly [2].

The amount of information in the world is increasing far more quickly than our ability to process it. The user has a lot many options to choose from RS helps in solving the excess information problem and helps the technical giants in generating good revenue by delivering actual value to their customers. RS helps the user in providing the item of their interest. Netflix recently generated a $1 billion revenue in 2020 as a result of proper RS implementation. The use of PDS helps in dealing with the huge data in a very efficient manner, and it also takes less time in performing operations on data.

11.5 PROPOSED WORK

A CF system creates embedding for both users and items on its own. It embeds both users and items in the embedding space. It considers other users' reactions while recommending a movie to a particular user. It finds out the common likings of a user and other users and considers them to recommend a movie. In this project, we are using CF to build a recommendation system enhanced using BF PDS. Using the BF, we are also considering the user's own choices besides the choices of similar users. To get better recommendations, we are adding the user's personal choices to three BFs signifying whether the person is interested in the movie/genre or not. After getting recommendations using CF, we query the BF to refine our recommendations to exclude the movies the user has already watched or is not interested in watching the movies. Since, the CF model implicitly assumes that if a user has watched a movie, it means he/she is interested in the movie, here we can conclude that a user is interested in a movie if he/she has watched the movie and given a positive review to it (rating ≥ 2.5), hence filtering out unnecessary movies.

11.5.1 INITIAL PROCESSING AND INSERTION PROCESS IN THE PROPOSED APPROACH

As shown in Algorithm 11.1, initially, we create a pivot table having movie title vs. userId values of ratings. Using this table, a correlation matrix is created, which gives how much appropriate it is to suggest another movie based on some movie taken from the table. For a particular userId, the topmost similar movies are found based on this correlation matrix and returned as a list.

Algorithm 11.1 Getting Recommendations from a Model Based on CF

1. *input_data* – data with fields: movieId, title, userId, rating
2. Create a pivot table *userRatings* from input_data such that columns correspond to movie title, rows correspond to userId, and values filled are ratings.
3. Get Recommendations for the user
   a. Create a correlation matrix *corrMatrix* consisting of correlations between ratings of different movies (based on Pearson Coefficient).
   b. Get *user_id* of the user for whom we need the recommendations
   c. **for all** *rows* in *input_data* with userId = user_id **do**
   d. *similarRatings <- corrMatrix[title] * (rating – 2.5)*
   e. *similarMovies <- similarRatings.sort(ascending = False)*
   f. **end for**
   g. *similarMovies* contains the recommendations with the best recommendation at the first place and the least recommended movie at the last.

11.5.2 QUERY PROCESS IN THE PROPOSED APPROACH

Out of the list of movies obtained from the normal RS to provide a better and relevant result to the user, all the recommendations are selected, and these recommendations are passed through three BFs for the user. The three BFs are *bloomc*, *bloomi*, and *bloomn*. *bloomc* stores the category of the movie liked by the user, *bloomn* stores the movie that the user didn't like these are those movies which have got the rating below 2.5 by the user, and *bloomi* stores the movie that is of user interest these are those movies which are rated ≥2.5 by the user. These BFs together return the category to which the recommendation belongs.

Algorithm 11.2 Refining the Recommendations Using a Bloom Filter (BF)

1. Create three Bloom Filters for each user: one for category/genre, another for movies in which the user is interested, and the last one for movies in which the user is not interested, namely *bloomc, bloomi, bloomn* respectively.

2. Get the results from Bloom Filters.
    a. *a, b, c* – the results from *bloomc, bloomn, bloomi* respectively.
    b. **if** *recommendation in bloomc* **then**
    c. **set** $a = 1$
    d. **end if**
    e. **if** *recommendation in bloomn* **then**
    f. **set** $b = 1$
    g. **end if**
    h. **if** *recommendation in bloomi* **then**
    i. **set** $c = 1$
    j. **end if**
3. Label the recommendations using the values of *a, b, c* obtained (See Table 11.1).

After looking at the categories corresponding to the values for *a, b, c*, the system can provide suggestions to the user. These categories help the system in providing a better decision than before. Betterment in suggestions is achieved by removing the redundancy and irrelevant suggestions like items with category wrong result, invalid is discarded from the list of recommendation. This way RS using BF helps in providing a suggestion that is better than before.

RS should be efficient because due to increase in the amount of data in the current time, users are very confused about how to get something useful and suitable for them, Therefore, RS is playing a vital role in helping the users in getting them something suitable for them for which the quality of RS implemented by the different platform should be met the expectations of the user. With the help of BF, it is becoming quite easy to have an RS of good quality and also BF takes less time and space to process the query and even time and space are valuable for doing any task.

As discussed in the above section, the proposed approach has shown a significant improvement in pruning the recommendation without adding any extra computational cost to the RS same has been tested and validated in the following section by considering some real-time datasets.

TABLE 11.1
Labels Given to Movies Based on Values of *a*, *b*, and *c* Calculated in Algorithm 11.2

a	b	c	Category
0	0	0	NEW
0	0	1	OLD INTEREST
0	1	0	WRONG RESULT
0	1	1	INVALID
1	0	0	HOT
1	0	1	OLD INTEREST
1	1	0	WRONG RESULT
1	1	1	INVALID

11.6 OBSERVATION AND ANALYSIS

All the experiments have been performed on the system having a 10th generation i5 processor. For writing python code for implementation, Anaconda navigator and Jupyter notebook have been used.

To use a BF, a BF library is used in the proposed solution. MovieLens 20M Dataset is used for this work. It is a stable benchmark dataset with 20 million ratings and 465,000 tag applications applied to 27,000 movies by 138,000 users It also includes tag genome data with 12 million relevance scores across 1,100 tags (See Figures 11.2 and 11.3).

Below are some snapshots of our experimental work for a better understanding that how our proposed model is working.

It can be observed that the CF-based RS returns a list of recommendations for a certain input user, identified by userId (see Figure 11.4). The output list consists of all the movies that the user might be interested to watch based on preferences of similar items. The list of results obtained from the RS is organized in a data frame and the resultant data (see Figure 11.5). This result is passed through the BF to label the recommendations into different categories.

	A	B	C	D
1	userId	movieId	rating	timestamp
2	1	2	3.5	02-04-2005 23:53
3	1	29	3.5	02-04-2005 23:31
4	1	32	3.5	02-04-2005 23:33
5	1	47	3.5	02-04-2005 23:32
6	1	50	3.5	02-04-2005 23:29
7	1	112	3.5	10-09-2004 03:09
8	1	151	4	10-09-2004 03:08
9	1	223	4	02-04-2005 23:46
10	1	253	4	02-04-2005 23:35

FIGURE 11.2 Some rows of the rating.csv file.

	A	B	C
1	movieId	title	genres
2	1	Toy Story (1995)	Adventure|Animation|Children|Comedy|Fantasy
3	2	Jumanji (1995)	Adventure|Children|Fantasy
4	3	Grumpier Old Men (1995)	Comedy|Romance
5	4	Waiting to Exhale (1995)	Comedy|Drama|Romance
6	5	Father of the Bride Part II (1995)	Comedy
7	6	Heat (1995)	Action|Crime|Thriller
8	7	Sabrina (1995)	Comedy|Romance
9	8	Tom and Huck (1995)	Adventure|Children
10	9	Sudden Death (1995)	Action

FIGURE 11.3 Some rows of the movie.csv file.

```
user = 15
recomends = giveResult(user)
genre=[]
movieIds=[]
finalResult=[]
for idx,row in movies.iterrows():
    if row['title'] in recomends:
        genre.append(row['genres'])
        movieIds.append(row['movieId'])
        a=(str(row['title']),row['genres'])
        finalResult.append(a)
print(finalResult)

[('American President, The (1995)', 'Comedy|Dram
lls (1995)', 'Comedy'), ('Copycat (1995)', 'Crim
er'), ('City of Lost Children, The (Cité des enf
(1995)', 'Children|Drama'), ('Clueless (1995)',
medy'), ('Beautiful Girls (1996)', 'Comedy|Drama
1 (1996)', 'Drama|Thriller'), ('Bottle Rocket (1
```

FIGURE 11.4   Code for getting recommendations for a particular user with userId = 15.

```
# Printing the obtained result
d = {'title':recomends,'Category':genre}
df = pd.DataFrame(d)
df
```

	title	Category
0	Ace Ventura: Pet Detective (1994)	Comedy|Drama|Romance
1	Ace Ventura: When Nature Calls (1995)	Crime|Drama
2	Adventures of Priscilla, Queen of the Desert, ...	Comedy
3	American President, The (1995)	Crime|Drama|Horror|Mystery|Thriller
4	Apollo 13 (1995)	Action|Crime|Thriller
5	Assassins (1995)	Adventure|Drama|Fantasy|Mystery|Sci-Fi
6	Babe (1995)	Children|Drama

FIGURE 11.5   Converting the result in the form of a dataframe.

The final list of recommendations from the RS is passed through BFs that are populated using the data. As explained in Algorithm 11.2, the recommendations are categorized into different labels. Based on these labels, we can filter out the recommendations that are not required.

11.7   RESULTS

This part of the paper shows how the use of BF in RS has improved the result in terms of time and space in handling huge data. It also minimizes the problem of irrelevant and redundant results by providing unique and useful decisions to the user and hence making the system better.

From Figure 11.6, it can be seen that the time taken before using BF for recommendation and after using BF is approximately the same, which implies that the new system provides better recommendations with no overhead of time.

It is evident that RS using BF removes the irrelevant and redundant suggestions and categorizes the recommendations into different categories proving the refinement in the suggestions (See Figure 11.7).

In some cases, the recommendations returned by both Normal RS and proposed system are same, indicating that the recommendations returned by Normal RS are

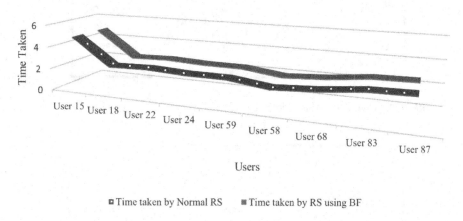

FIGURE 11.6   Time taken by Normal RS vs. Time taken by RS using BF for a random sample of users.

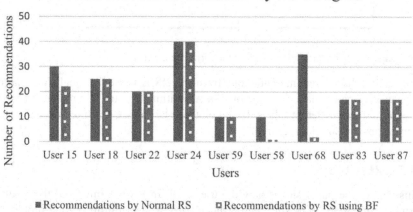

FIGURE 11.7   Number of recommendations returned by Normal RS vs. Number of recommendations returned by RS using BF for a random sample of users.

all relevant to the user taken into consideration. Without taking any extra time, the recommendations are double-checked using the BF and hence, the adequacy of the recommendations is also verified.

Thus, it can be said that the recommendations are becoming better than before all because of using the BF in the existing RS.

11.8 CONCLUSION AND FUTURE SCOPE

The research focuses on pruning in RS using learned BF. A brief study of BF makes us realize its importance in the field of Big Data, and using BF, which is a PDS for handling massive data, makes the operation easy, besides being time and space efficient. An overview of the existing implementation of BF in RS is illustrated. The major challenge associated with Normal RS includes irrelevant, redundant, overspecialization, scalability, and a huge amount of search time. The experimental results outperform in terms of providing better recommendations to the user by removing redundancy and irrelevant suggestions for the user and using BF which takes constant time for operating on tremendous data.

Using a BF improves the results of an RS without any overhead of time complexity. The list of recommendations provided bypassing the results (from RSs) through a BF is refined to provide the user with recommendations suited to his/her choice along with other users' choices.

In the future, one can focus on context-aware RS in the future course and discuss different efficient approaches to context selection, as well as other future developments in RS, to make RS more reliable and realistic to use. Indirect feedback for objects is being implemented. Other BF, such as content-based filtering and dynamic BFs, will be tested in the future. Both activities are critical in RS to enhance recommendations and make them more scalable.

REFERENCES

1. Pagare, R. & Shinde, A. (2013). Recommendation system using bloom filter in mapreduce. *International Journal of Data Mining & Knowledge Management Process*, 3, 127–134. doi: 10.5121/ijdkp.2013.3608.
2. Vaidya, N. & Khachane, A. Recommender systems-the need of the e-commerce ERA. In *2017 International Conference on Computing Methodologies and Communication (ICCMC)*, 2017, pp. 100–104. doi: 10.1109/ICCMC.2017.8282616.
3. Khusro, S., Ali, Z., & Ullah, I. Recommender systems: Issues, challenges, and research opportunities. In *Information Science and Applications (ICISA)*, 2016. doi: 10.1007/978-981-10-0557-2_112.
4. Anandhan, A., Shuib, L., Ismail, M. A., & Mujtaba, G. (2018). Social media recommender systems: Review and open research issues. *IEEE Access*, 6, 15608–15628. doi: 10.1109/ACCESS.2018.2810062.
5. Lee, J., Sun, M., & Lebanon, G. (2012). A comparative study of collaborative filtering algorithm. arXiv preprint arXiv:1205.3193.
6. Sarwar, B., Karypis, G., Konstan, J., & Riedl, J. Item-based collaborative filtering recommendation algorithms. In *Proceedings of ACM World Wide Web Conference*, 2001. 1. doi: 10.1145/371920.372071.

7. Rohani, V. A., Kasirun, Z. M., & Ratnavelu, K. An enhanced content-based recommender system for academic social networks. In *2014 IEEE Fourth International Conference on Big Data and Cloud Computing*, 2014, pp. 424–431. doi: 10.1109/BDCloud.2014.131.

8. Dou, Y., Yang, H., & Deng, X. A survey of collaborative filtering algorithms for social recommender systems. In *2016 12th International Conference on Semantics, Knowledge, and Grids (SKG)*, 2016, pp. 40–46. doi: 10.1109/SKG.2016.014.

9. Zhang, R., Liu, Q., Chun-Gui, J. W., & Huiyi-Ma. Collaborative filtering for recommender systems. In *2014 Second International Conference on Advanced Cloud and Big Data*, 2014, pp. 301–308, doi: 10.1109/CBD.2014.47.

10. Wang, Y., Chan, S. C., & Ngai, G. Applicability of demographic recommender system to tourist attractions: A case study on trip advisor. *2012 IEEE/WIC/ACM International Conferences on Web Intelligence and Intelligent Agent Technology*, 2012, pp. 97–101, doi: 10.1109/WI-IAT.2012.133.

11. Chen, H., Li, X., & Huang, Z. Link prediction approach to collaborative filtering. In *Proceedings of the 5th ACM/IEEE-CS Joint Conference on Digital Libraries (JCDL '05)*, 2005, pp. 141–142. doi: 10.1145/1065385.1065415.

12. Kluver, D., Ekstrand, M., & Konstan, J. (2018). Rating-based collaborative filtering: Algorithms and evaluation. doi: 10.1007/978-3-319-90092-6_10.

13. Cacheda, F., Carneiro, V., Fernández, D., & Formoso, V. (2011). Comparison of collaborative filtering algorithms: Limitations of current techniques and proposals for scalable, high-performance recommender systems. *WEB*, 5, 2.

14. Iaquinta, L., de Gemmis, M., Lops, P., Semeraro, G., Filannino, M., & Molino, P. Introducing serendipity in a content-based recommender system. In *Proceedings -8th International Conference on Hybrid Intelligent Systems, HIS 2008*, 2008, 168–173. doi: 10.1109/HIS.2008.25.

15. Su, X. & Khoshgoftaar, T. (2009). A survey of collaborative filtering techniques. *Advances in Artificial Intelligence*, 1, 1–9. doi: 10.1155/2009/421425.

16. Linden, G., Smith, B., & York, J. (2003). Amazon.com recommendations: Item-to-item collaborative filtering. *IEEE Internet Computing*, 7(1), 76–80. doi: 10.1109/MIC.2003.1167344.

17. Mooney, R. J. & Roy, L. Content-based book recommending using learning for text categorization. In *Proceedings of the Fifth ACM Conference on Digital Libraries (DL '00)*, 2000. Association for Computing Machinery, New York, NY, USA, pp. 195–204. doi: 10.1145/336597.336662.

18. Quijano-Sanchez, L., Cantador, I., Cortés-Cediel, M., & Gil, O. (2020). Recommender systems for smart cities. *Information Systems*, 92, 101545. doi: 10.1016/j.is.2020.101545.

19. Shi, Y., Larson, M., & Hanjalic, A. (2014). Collaborative filtering beyond the user-item matrix: A survey of the state of the art and future challenges. *ACM Computing Surveys*, 47(1), 1–45. doi: 10.1145/2556270.

20 Karabadji, N. E. I. et al. (2018). Improving memory-based user collaborative filtering with evolutionary multi-objective optimization. *Expert Systems with Applications*, 98, 153–165.

21. Chen, H. & Li, J. Learning multiple similarities of users and items in recommender systems. In *2017 IEEE International Conference on Data Mining (ICDM)*, 2017, pp. 811–816.

22. Singh, A., Garg, S., Kaur, R., Batra, S., Kumar, N., & Zomaya, A. (2019). Probabilistic data structures for big data analytics: A comprehensive review. *Knowledge-Based Systems*, 188, 104987. doi: 10.1016/j.knosys.2019.104987.

23. Pozo, M., Chiky, R., Meziane, F., & Métais, E. An item/user representation for recommender systems based on bloom filters. In *2016 IEEE Tenth International Conference on Research Challenges in Information Science (RCIS)*, 2016, pp. 1–12. doi: 10.1109/RCIS.2016.7549311.

12 Performance Evaluation of ARIMA and CRNN for Weather Prediction

Uma Sharma and Chilka Sharma
Banasthali Vidyapith

CONTENTS

12.1 INTRODUCTION

Weather conditions change rapidly and continuously across the globe. Accurate forecasting is required to be essential in multiple aspects related to human survival. In the field of agriculture, industry, travelling, and military surveillance, we are reliant a lot on forecasting of weather. Because the entire world is grieving from the incessant change of climate as well as its side effects also, it is much required to predict the weather accurately to ensure easier and flawless mobility, along with secure daily-based operations [1–4].

Research community is getting more and more interested in weather prediction, mainly because of its various impacts on human survival. A number of lives can possibly be saved with the prior warning of potential climate disasters over the years to come [5–7]. Prediction of the climate conditions has been proving its importance

DOI: 10.1201/9781003320333-12

in the agricultural and farming sectors, such as storage of food grains for future use and transportation of food grains and various agricultural products to various areas [8,9]. Prior information of multiple types of calamities like tornados, cyclones, and heavy rainfall has proved to be of very much importance for the safety and protection of human life [4]. Further damage incurred as a national loss can be reduced with the help of such kind of predictions [10]. Transportation and service sectors are also hampered due to adverse weather conditions, resulting in an increase in various risks like accidents and also delays in the offered service as well as false impacts on the quality of service [11–13]. Accuracy in weather prediction has proven to be useful in discovering the right decisions and also helps in taking much-needed wise actions. That is why increasing the efficiency of weather forecasting provides huge benefits in various long- and short-term active plans [14–17].

In geosciences, applying machine learning models to meteorological data brings various opportunities, such as faithful prediction of future climate conditions. In recent years, the use of deep neural networks in modeling meteorological data has become a significant area of research [18]. The aim of the research is to find a method that enhances the forecasting results, related to different parameters. It is desirable to implement such methods for forecasting and match the outcome in the prediction of various parameters such as temperature, air pressure, humidity, and wind speed. Meteoric knowledge on the other hand is considered to be tentative (unsure) in nature, and knowledge on weather is mostly outlined. It is well-known that weather knowledge will have the noises and outliers, and this is the main reason for incorrect analysis. Noises are the arbitrary error; therefore, to improve the criterion of knowledge, prepossessing of weather knowledge is required for accurate and precise weather prediction. In this work, two automatic forecasting methods, (autoregressive integrated moving average (ARIMA) and convolutional recurrent neural network (CRNN), are analyzed and compared on the scale of various weather parametric data like temperature, humidity, wind speed, wind bearing, visibility, and air pressure. A relative study was done between ARIMA and CRNN, and the performance of both models was compared by using various regression metrics, such as MSE, RMSE, MAE, and R-squared error. [19].

Some more models will be evaluated in the future works, such as using deep learning and neural networks besides long short-term memory (LSTM).

This chapter follows the following structure: Section two goes through related work; Section three describes methodology, which includes the data set, it's preprocessing and forecasting models; Section four illustrates experimental results and analysis of ARIMA model and deep learning model CRNN; and lastly, Section five discusses the conclusion and future work.

12.2   RELATED WORK

A research was conducted by Mark Hallstrom, Dylan Liu, and Christopher Vo to exert machine learning to predict climate [20]. The illustrated technique in this study uses data collected from weather underground. The collected data set includes minimum temperature, maximum temperature, and mean pressure of the atmosphere, mean humidity, and daily weather condition during the years of 2011–2015 in Stanford. This study is constricted to predict minimum and the maximum temperature for seven days ahead by using the data from the past two days. Four weather classifications, which are namely rain, precipitation, cloudy, and very cloudy, were used in this study. Linear regression algorithm was used to forecast maximum and minimum temperature as a linear combination. Further, a variation of functional algorithms was used to identify previous weather conditions that were analogous to the current weather condition. Then, these past weather conditions were used to predict the weather.

Moreover, this study has tested functional regression and linear regression models with few days. Further, both professional weather forecasting services were able to outperform both the of machine learning models that were discussed in the study. It was found that the professional weather forecasting services' performances decreased over time. As for the future improvements, this study proposes to collect more data to increase the performance of the linear regression models.

A research was done by Tarun Rao, N. Rajasekhar, and T. V Rajinikanth to forecast weather using support vector machines (SVMs) as a bright approach [21]. The algorithms of SVM fall into the supervised machine learning model category, and they're widely used for classification and regression analysis problems. In this research, linear support vector regression is utilized to predict maximum temperature of the day. Further, in this study, multi-layer perceptrons, which are trained with back propagation method, are applied to compare with the SVM model to predict weather. Perceptrons, which are widely known as neural networks, also mostly fall into the supervised machine learning algorithm category. Neural networks mimic the way animal and humans learn in order to develop a learning model. Back propagation is the method that is used to fine-tune the weights of each neuron in the network by backward propagating the error to inner layers of the network. To train each of these models, this study has used a data set from the University of Cambridge, which has a span of 5 years.

According to the results, the utilization of SVMs for weather forecasting was able to outperform the multi-layer perceptron neural network, which is trained by the use of back propagation algorithm. The only drawback is that training time of SVMs is increased when dealing with higher orders.

A study was done by Lai et al. on weather forecasting using dynamically weighted time delay neural networks in the east china [22]. Data from fourteen local weather centers during 1991 to 2001 time period is used to train and validate the dynamic weighted time delay neural network used in the study. Dynamic weighted time delay neural network is a subcategory of neural networks. Time delay neural networks have the ability to identify translation invariant shapes and model the contest at every layer.

According to the results of this study, the author has come to a conclusion that an even neural network with a single hidden layer can approximately predict the rainfall and temperature.

A study was done by Warangkhana Kimpan and Saktaya Suksri, which describes a neural network model for weather prediction using the fireworks algorithm [23]. In this study, the neural network is trained via the fireworks algorithm. This algorithm is a swarm intelligence algorithm, which mimics the behavior of different groups.

Fireworks algorithm is mainly used for optimization purposes. The aim is to forecast the mean temperature of a day. This study uses past daily weather between 2012 and 2015 from Chalermprakiet weather station as the data set. The model attained training accuracy of 81.48% and testing accuracy of 73.79%. This study suggests comparison with other prediction algorithms as for the future works.

ARIMA model was used in the experiments done by the authors [24] for visibility forecasting and resulting lowest MSE values, i.e., 0.00029 and also suggests to use LSTM with convolutional neural network (CNN) for better results for future work. In the work done by the authors [25] for short-term weather forecasting up to 12 hours, various lightweight deep learning models like LSTM, temporal convolutional network used for ten surface parameters and give better results than the state-of-art WRF model. In a comparison done by Yuhu Zhang et al. [26] for drought forecasting using ARIMA, wavelet neural network, and SVM, it shows that ARIMA was advantageous than other two models in terms of Nash–Sutcliffe efficiency coefficient (NSE) and R^2.

12.3 METHODOLOGY

For the purpose of weather forecasting, some essential parameters are considered, such as time, summary, precipType, temperature, apparent temperature, humidity, wind speed, wind bearing, visibility, loudCover, and pressure.

As shown in Figure 12.1, the data collected for research is different in nature and may contain some useless information or missing values, so to convert these inaccurate values to standard data, we will first do preprocessing of that data using the standard scaling technique.

After this, we will have accurate and desired data; then, we can apply different methods or algorithms for prediction of future weather and have weather condition alerts. Next section will describe preprocessing details followed by the methods used for prediction of weather data.

12.3.1 Data set and Preprocessing

Weather data from Szeged, Hungary has been used for this study. Data set contains hourly weather data, and total instances are (96452,11); the attributes in the data set are time, summary, precipType, temperature, apparent temperature, humidity, wind speed, wind bearing, visibility, loudCover, and pressure. Data set covers the time period between 2006 and 2016.

Data was generally inconsistent, inaccurate, and had missing values, so we had to first clean the data to make it accurate for our use, and for this, a method is used, i.e., to replace the missing or noisy values by forward filling them using previous points in time; for example, missing temperature value is filled with the value of the last recorded temperature. After cleaning the data, all of the values were normalized to points in [1; 1] to neglect training problems and also weight decay.

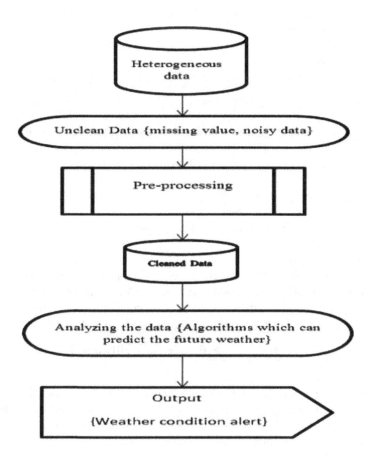

FIGURE 12.1   Procedure of proposed model.

12.3.1.1   Standard Scaling-Based Preprocessing

Standard scaling is a preprocessing-based scaling method in which the values are centered on the mean with a unit standard deviation. As a resultant, the mean of the attribute becomes zero and the resultant distribution has a unit standard deviation. Following is the formula for standardization:

$$X' = \frac{X - \mu}{\sigma} \tag{12.1}$$

where σ is the standard deviation of the feature values, and μ is the mean of the feature values. It's worth noting that the numbers in this situation aren't limited to a specific range.

12.3.2 FORECASTING MODELS

12.3.2.1 ARIMA

According to research conducted by Box and Jenkins [27], ARIMA ((p, d, q) Auto regressive integrated moving average model), which is a part of stochastic processes, was used to analyze time series data. Since then, ARIMA models have been widely applies in a wide range of time series analysis applications. A research was done by Ling Chen and Xu Lai [28], which involves wind speed forecasting by applying ARIMA model methodology of box and Jenkins as described.

Step 1: Model recognition
Step 2: Estimation of parameters
Step 3: Checking the residual diagnostics and forecasting

Step1: Model recognition
ARIMA models are one of the most extensively used approaches for time series forecasting. For a time series which is stationary, ARIMA (p, d, q) model can be written in terms of past temperature data, residuals, and prediction errors as follows:

$$x_t = \sum_{i=1}^{p} \varphi_i x_{t-i} - \sum_{j=1}^{q} \theta_j a_{t-j} + a_t a_t \sim \text{NID}\left(o, \sigma^2\right) \tag{12.2}$$

Temperature time series data is denoted by x, and $(t-i)$th data is denoted by x_{t-i}, random white noise time series is denoted by at, auto regressive parameters are denoted by φ_i, moving average parameters are denoted by θ_j, order of auto regressive model is denoted by p, order of moving average model is denoted by q and degree of differencing is denoted by d.

Dickey–Fuller test can be used to check whether the given time series is stationary or not. Augmented Dickey–Fuller tests are generalized Dickey–Fuller tests, which can accommodate ARIMA models.

If the time series is nonstationary differencing is performed to transform time series into a stationary model. First-order differencing can be expressed as below,

$$y_t = x_{t+1} - x_t \tag{12.3}$$

If y_t isn't stationary, when $d=1$ then, it is needed to difference using $d-1$ times till it becomes stationary.

Step2: Estimation of parameters
Parameter estimation involves choosing the right p and q, which can describe the time series with the highest accuracy. These parameters can be approximately estimated by analyzing the autocorrelation function (ACF) and the partial autocorrelation function (PACF). The ACF function can be helpful to determine the order

of moving average q, and PACF function can be used to determine the order of the auto regressive model. P value is the lag value where PACF graph crosses the upper confidence gap. Q value is the lag value where ACF graph crosses the upper confidence gap.

Step3: Checking the residual diagnostics
Residuals are leftovers after fitting an appropriate model. In most cases, residuals can be regarded as the difference between the original values and the forecasted values.

$$a_t = y_{t-}\hat{y}_t \qquad (12.4)$$

Residual diagnostic is performed to check whether the model has been able to capture adequate information in the time series data. For an appropriate forecasting model, residuals should be uncorrelated and should have zero mean.

12.3.2.2   Deep Learning Model CRNN
Deep learning model discussed in this study consists of a one-dimensional convolutional layer and two LSTM layers. One-dimensional convolutional layers are used to obtain lateral features from the time series data [29]. LSTM layers are used to extract temporal features from the temperature time series. High-level architecture of the deep learning model is shown in Figure 12.2.

12.3.2.2.1   One-Dimensional Convolution Layer
CNNs have evolved in recent years. Convolution operation is a function, which is derived from two given functions where it can be used express how the shape of one function can be modified by the other function. CNNs are trained using back propagation algorithm [30]. Furthermore, CNNs are commonly used for image processing-related tasks. CNNs share similar properties and similar approach without considering the dimensionality. But the main difference is from the input data dimensionality and the way filter slides across the given data. In a one-dimensional convolutional layer filter slides along a single dimension. Our temperature time series data set has two dimensions; they're time steps and the temperature values. Therefore, filter or kernel can only move along the time dimension (Figure 12.3).

Generally, pooling is applied to the filtered resulting image. The purpose of pooling is to achieve local translation invariance. Pooling achieves this purpose by down

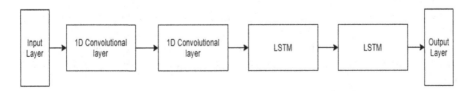

FIGURE 12.2   High level architecture of the deep learning model.

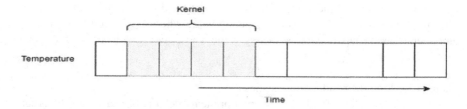

FIGURE 12.3 Kernel sliding over temperature data.

sampling the feature maps. There are two pooling methods. They're average pooling and max pooling. Initially, feature maps are divided into chunks. In max pooling, maximum value of a group is taken. In average pooling, average value of the group is taken. The resulting image is a shrunken version of the original.

In this work, to find out best possible forecasting approach, a 1D CNN is used, which comprises of one convolutional layer along with two fully connected layers. Following settings are been used in the convolutional layer: 64 filters, stride equal to 1, kernel size equal to 3, padding is set to "same," and "relu" is used as activation function for 200 epochs.

12.3.2.2.2   LSTM Layer

According to a research paper by Hochreiter and Schmidhuber, long short-term memory networks (LSTMs) were first introduced to the world [30]. LSTMs are a special type of RNNs (recurrent neural networks). LSTMs have the ability of learning from long-term dependencies. RNNs use truncated back propagation through time for training. But, RNNs suffer from vanishing gradient problem when it has too many time steps. LSTMs are designed to overcome vanishing gradient problem. LSTMs update its cell state and various gates (Figure 12.4).

Forget gate: Forget gates' purpose is to decide which information should forget from the memory. For that, sigmoid function is applied to information from the previous hidden state and the current input. If the value is close to zero means abandon and closer to 1 means retain.

Input Gate: Input gate concerned with updating the cell state. It first decides which values should be updated by applying sigmoid function to the previous hidden state and current input, where 0 means no updates necessary and 1 means should be updated. Then, a tanh gate is applied to transform values between −1 and +1. Next outputs from these two gates are multiplied and added to the cell state.

Output gate: Output gate involves deciding the next hidden state. A sigmoid function is applied to the previous hidden state and the current input, to decide which information should contain in the next hidden state. After that, tanh gate is applied to the new cell state. Next hidden state is calculated multiplying these two gate values [31].

In this work, an LSTM network is used which consist of two stacked LSTM layers, out of which one layer is of 32 LSTM units and other is of 16 LSTM units, along with two fully connected layers. The number of neurons is set to 10 in the first fully

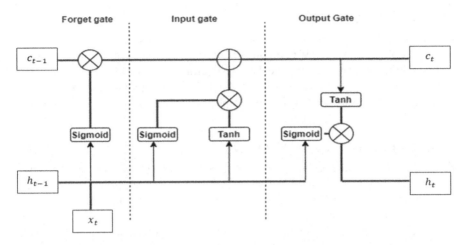

FIGURE 12.4 LSTM cell structure.

TABLE 12.1
Details of Different Evaluation Metrics Used

MSE / RSME	MAE	R²
Based on square of error	Based on absolute value of error	Based on correlation between actual and predicted value
Value lies between zero to infinity	Value lies between zero to infinity	Value lies between 0 and 1
Perceptive to outliers, punishes larger error more	Treat larger and small errors equally. Not sensitive to outliers	Not perceptive to outliers
Small value indicates better model	Small value indicates better model	Value near 1 indicates better model

connected layers and equal to the number of output variables in the second fully connected layer (output layer), and activation function used is "relu" for 200 epochs.

12.4 EXPERIMENTAL RESULTS AND ANALYSIS

To evaluate the proposed ARIMA and deep learning model CRNN, same preprocessed data set is used for identifying ARIMA model, for training the deep learning model and testing the both of these models. After that for performance evaluation of each of the methods, we have used different parametric evaluation metrics, such as MSE, RMSE, MAE, and R-squared error (Table 12.1).

12.4.1 RESULTS

In this section, results for both models that is ARIMA and deep learning model CRNN are displayed individually.

12.4.1.1 ARIMA

In our model for ARIMA (p, d, q), we have taken $p=5$, $d=1$, and $q=0$, and total number of observations is 7,976. Figure 12.5 displays the result of ARIMA model.

Figure 12.6 displays the graphical analysis of actual temperature and predicted temperature for ARIMA model in which x-axis is for time(day) and y-axis is daily temperature in Celsius, and finally, the graph shows that both actual and predicted temperature are almost matches.

12.4.1.2 Deep Learning Model CRNN

Figure 12.7 displays the graphical analysis of actual temperature and predicted temperature for deep learning CRNN model in which x-axis is for time(day) and y-axis is daily temperature in Celsius, and finally, the graph shows that both actual and predicted temperature are almost matches.

The above graph in Figure 12.8 gives the percentage values for the performance metrics for deep learning model CRNN for various epoch levels till 200 epochs, in which the x-axis is the epoch range and the y-axis is the percentage level.

12.4.1.3 Analysis between ARIMA and Deep Learning Model CRNN

The numerical results on Table 12.2 describe the comparative analysis of performance metrics between ARIMA and deep learning model CRNN. According to the

```
                             ARIMA Model Results
==============================================================================
Dep. Variable:                    D.y   No. Observations:               7976
Model:                 ARIMA(5, 1, 0)   Log Likelihood            -17316.814
Method:                       css-mle   S.D. of innovations            2.122
Date:                Sun, 08 Aug 2021   AIC                        34647.629
Time:                        18:48:59   BIC                        34696.518
Sample:                             1   HQIC                       34664.365

==============================================================================
                 coef    std err          z      P>|z|      [0.025      0.975]
------------------------------------------------------------------------------
const          0.0001      0.018      0.007      0.994      -0.035       0.035
ar.L1.D.y      0.1119      0.011     10.001      0.000       0.090       0.134
ar.L2.D.y     -0.2087      0.011    -18.629      0.000      -0.231      -0.187
ar.L3.D.y     -0.0866      0.011     -7.595      0.000      -0.109      -0.064
ar.L4.D.y     -0.0966      0.011     -8.621      0.000      -0.119      -0.075
ar.L5.D.y     -0.0489      0.011     -4.374      0.000      -0.071      -0.027
                                     Roots
==============================================================================
                 Real          Imaginary           Modulus         Frequency
------------------------------------------------------------------------------
AR.1           0.9621           -1.1302j            1.4842           -0.1378
AR.2           0.9621           +1.1302j            1.4842            0.1378
AR.3          -0.6725           -1.7840j            1.9065           -0.3074
AR.4          -0.6725           +1.7840j            1.9065            0.3074
AR.5          -2.5536           -0.0000j            2.5536           -0.5000
```

FIGURE 12.5   ARIMA model results.

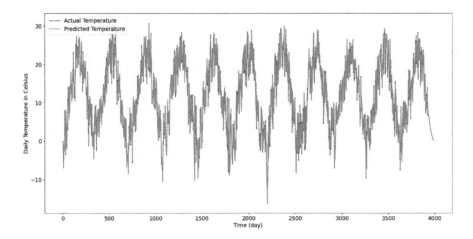

FIGURE 12.6   Graphical analysis of the actual and the predicted temperature for ARIMA model.

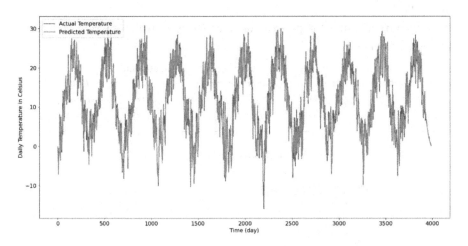

FIGURE 12.7   Graphical analysis of the actual and the predicted temperature at epoch 200 for deep learning model CRNN.

results, it can be clearly seen that deep learning model CRNN is giving superior performance than the ARIMA model because the error is less and value of R-squared is comparatively large for CRNN.

Figure 12.9 displays the graphical analysis of overall performance evaluation of ARIMA and CRNN for different metrics, and results prove that CRNN is better than ARIMA in all aspects.

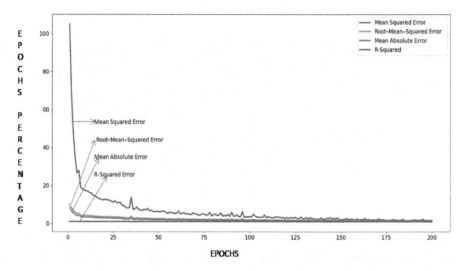

FIGURE 12.8 Percentage values for the performance metrics for various epoch levels for deep learning model CRNN.

TABLE 12.2
Overall comparison of the performance measures between ARIMA and deep learning model CRNN

Models/Metrics	Mean squared error	Root-mean-squared error	Mean absolute error	R-squared
ARIMA	2.252560693	1.500853322	0.806817394	0.470462427
CRNN	1.492859132	1.221826146	0.771068043	0.980424308

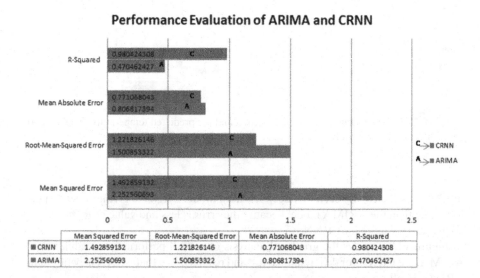

FIGURE 12.9 Performance evaluation of ARIMA and CRNN.

12.5   CONCLUSION AND FUTURE WORK

This chapter presented an analysis of two weather forecasting models, ARIMA (which is statistical analysis model that uses time series data) and CRNN (deep learning model).

The results are obtained by using similar data set on the scale of 200 epochs, which clearly shows that CRNN performs better than ARIMA. Papers were reviewed and analyzed to emphasize the capabilities of ARIMA and CRNN in the prediction of numerous weather phenomena such as temperature, humidity, wind speed, and atmospheric pressure. The prediction abilities of both the methods have been analyzed on certain performance metrics like MSE, RMSE, MAE, and R-squared error. The future study will be focused on the use of deep learning and statistical analysis model for some more aspects like prediction of more meteorological attributes along with consideration of their correlation, so as to improve the accuracy or timeliness.

REFERENCES

1. P. Nurmi, A. Perrels, and V. Nurmi, "Expected impacts and value of improvements in weather forecasting on the road transport sector," *Meteorol. Appl.*, vol. 20, no. 2, pp. 217–223, 2013.
2. V. Nurmi et al., "Economic value of weather forecasts on transportation–Impacts of weather forecast quality development to the economic effects of severe weather," no. 233919, p. 92, 2012.
3. WMO, "Guide to public weather services practices," no. 834, 1999.
4. World Bank, "Cities and climate change an urgent agenda," *Urban Dev. Local Gov.*, vol. 10, p. 279, 2010.
5. A. Giddens, "The politics of climate change," *Policy Polit.*, vol. 43, no. 2, pp. 155–162, 2015.
6. A. Costello et al., "Managing the health effects of climate change. Lancet and University College London Institute for Global Health Commission," *Lancet*, vol. 373, no. 9676, pp. 1693–1733, 2009.
7. A. Dupont, "The strategic implications of climate change," *Survival (Lond).*, vol. 50, no. 3, pp. 29–54, 2008.
8. D. Bergholt and P. Lujala, "Climate-related natural disasters, economic growth, and armed civil conflict," *J. Peace Res.*, vol. 49, no. 1, pp. 147–162, 2012.
9. R. Sah, "Priorities of developing countries in weather and climate," *World Dev.*, vol. 7, no. 3, pp. 337–347, 1979.
10. J. S. Pender, "What is climate change? and how it may affect Bangladesh," *Nor. Church Aid*, no. June, pp. 1–75, 2008.
11. J. Pender, "Climate change, its impact and possible community based responses in Bangladesh," *Change*, pp. 1–97, 2010.
12. Food and Agriculture Organization of the United Nations (FAO), The Impact of disasters and crises on agriculture and Food Security. 2017.
13. P. Leviäkangas et al., Extreme weather impacts on transport systems. EWENT Project Deliverable D1. 2011.
14. World Bank - Global Facility for Disaster Reduction and Recovery (GFDRR), "Background paper on the benefits and costs of early warning systems for major natural hazards," 2009.
15. L. Goddard et al., "Providing seasonal-to-interannual climate information for risk management and decision-making," *Procedia Environ. Sci.*, vol. 1, no. 1, pp. 81–101, 2010.

16. G. Ziervogel, P. Johnston, M. Matthew, and P. Mukheibir, "Using climate information for supporting climate change adaptation in water resource management in South Africa," *Clim. Change*, vol. 103, no. 3, pp. 537–554, 2010.

17. U. Sharma, and C. Sharma, "Analysis and literature review of weather prediction and forecasting methods," *Test Eng. Manage.*, vol. 83, no. May – June, pp. 12539–12549, 2020.

18. M. Tektas, "Weather forecasting using ANFIS and ARIMA, "A case study for Istanbul," *Environ. Res. Eng. Manage.*, vol. 1, no. 51, pp. 5–10, 2010.

19. G. Jain, and M. Bhawna, "A study of time series models ARIMA and ETS." Available at SSRN 2898968 (2017).

20. M. Holmstrom, D. Liu and C. Vo, "Machine learning applied to weather forecasting," *Meteorol. Appl.*, vol. 10, pp. 1–5, 2016.

21. T. Rao, N. Rajasekhar and T. V. Rajinikanth, "An efficient approach for Weather forecasting using Support Vector Machines," *IPCSIT*, vol. 47, pp. 208–212, 2012.

22. L. L. Lai, et al., "Intelligent weather forecast." *Proceedings of 2004 International Conference on Machine Learning and Cybernetics (IEEE Cat. No. 04EX826)*, vol. 7, pp. 4216–4221. IEEE, 2004.

23. S. Suksri and W. Kimpan, "Neural network training model for weather." *International Computer Science and Engineering Conference (ICSEC)*, pp. 1–7, 2016.

24. A. G. Salman, and B. Kanigoro. "Visibility forecasting using Autoregressive Integrated Moving Average (ARIMA) Models." *Procedia Comput. Sci.*, vol. 179, no. 2021, pp. 252–259, 2021.

25. P. Hewage, M. Trovati, E. Pereira, and A. Behera. "Deep learning-based effective fine-grained weather forecasting model." *Pattern Anal. Appl.*, vol. 24, no. 1, pp. 343–366, 2021.

26. Y. Zhang, H. Yang, H. Cui, and Q. Chen. "Comparison of the ability of ARIMA, WNN and SVM models for drought forecasting in the Sanjiang Plain, China." *Nat. Resour. Res.*, vol. 29, no. 2, pp. 1447–1464, 2020.

27. G. E. Box, G. M. Jenkins, G. C. Reinsel, and G. M. Ljung, "Time series analysis: forecasting and control," 1976.

28. L. Chen, and X. Lai. "Comparison between ARIMA and ANN models used in short-term wind speed forecasting." *2011 Asia-Pacific Power and Energy Engineering Conference*, pp. 1–4. IEEE, 2011.

29. Y. LeCun, L. Bottou, Y. Bengio, and P. Haffner, "Gradient based learning applied to document," *Proc. IEEE*, vol. 86, no. 11, 2278–2324, 1998.

30. S. Hochreiter, and J. Schmidhuber, "Long short-term memory," *Neural Comput.*, vol. 9, no. 8, 1735–1780, 1997.

31. E. De Saa, and L. Ranathunga. "Comparison between ARIMA and deep learning models for temperature forecasting." arXiv preprint arXiv:2011.04452 (2020).

13 A Novel Approach to Sorting Algorithm

Rushank Shah, Riddham Gadia, and Abhijit Joshi
Dwarkadas J. Sanghvi College of Engineering

CONTENTS

13.1 INTRODUCTION

The main aim of a sorting algorithm is to sort a given set of numbers in either ascending or descending form by following a fixed series of operations or steps. There are two types of sorting algorithms in general. The first one is comparison-based [1] sorting. Examples of comparison-based sorting techniques are insertion sort, selection-based sorting, quick sort, etc. The second type is non-comparison-based sorting. Examples of non-comparison-based sorting techniques are bucket sort, radix sort, etc. The proposed algorithm falls in the second category, where the algorithm uses hashing technique to sort the elements.

The algorithm proposed in this chapter uses hashes to sort a given list of numbers. Every element is indexed in a new array, and the index of this element is the value of the element. A further explanation of the proposed algorithm is given in detail in Section 13.3.

DOI: 10.1201/9781003320333-13

179

13.2   RELATED WORKS

Since the 1950s, research has been going on for different techniques to implement and improve sorting algorithms and techniques. With time, hardware has improved drastically, and there is visually no difference for the times taken for the different algorithms. However, when there is a limitation in the amount of computation power available, efficient algorithms always give us an advantage. The three main criteria for comparing algorithms are time complexity, space complexity [2], and stability [3].

The time complexity of any algorithm is reviewed by three cases, namely, "worst case", "average case", and "best case". Asymptotic notations [4] are used to denote the time and space complexities of an algorithm. The most used standard asymptotic function is the big-O notation [5]. It denotes the upper bound of a function. This tells us the maximum number of resources required by an algorithm to execute.

The stability of an algorithm is the ability of an algorithm to maintain the relative order of the elements with equal values. It is used to classify different sorting algorithms based on their relative positions before and after the algorithm is performed over them.

There are a plethora of sorting algorithms in existence [6,7]. This section discusses a few of them in brief.

The first algorithm is bubble sort [8]. It is the most primitive algorithm and is the slowest sorting algorithm [9]. It works by comparing each element to its neighboring element and swaps them if the order is incorrect. This process is continued until all the elements are sorted and the list is in the correct order. The worst-case and average-case complexities of this algorithm are $O(n^2)$, and the best-case complexity is $O(n)$.

The second algorithm is insertion sort [10]. It is a very efficient algorithm [11], and it is also simple. It performs very efficiently on small and almost sorted arrays [9]. We insert elements based upon their appropriate positions in the sorted list. Every element is sent to its appropriate position by comparing it to neighboring elements. It is an in-place algorithm. The worst-case and average-case complexities of this algorithm are $O(n^2)$, and the best-case complexity is $O(n)$. The worst-case and average-case complexities of this algorithm are $O(n^2)$, and the best-case complexity is $O(n)$.

The third algorithm is selection sort [12]. This algorithm performs better than bubble sort [9] but worse than insertion sort [11]. It works by searching for the least (or greater depending on the use case) element in the list and then swapping it with the second element in the sorted list. Every iteration will sort the start of the list and, at the end, sort the whole list. Selection sort is an in-place algorithm. The time complexity of this algorithm is $O(n^2)$ in all cases.

The fourth algorithm is heap sort [13]. It is a type of sorting, which uses heap data structure to sort the number. A heap is a tree data structure that satisfies the heap property. Heap sort stores the numbers in the heap and then removes the maximum element from the heap and inserts it at the end of the final sorted array, i.e., $n-1$st position. Every time, it is made sure that the heap property is not violated. The process is continued until the array is sorted. Heap sort is not a stable sort. The time complexity of this algorithm is $O(n\log(n))$ in all cases.

The fifth algorithm is merge sort [14]. It is an algorithm that is good for larger data set [9]. It follows a divide-and-conquer approach. It divides the array into smaller parts until the size of the sub-array becomes one. Then, a merge function is called to merge individual sub-array, and a final array is returned. This process is carried out recursively and continues until the entire array is sorted. Merge sort is a stable sort. The time complexity of this algorithm is $O(n\log(n))$ in all cases.

The sixth algorithm is quick sort [15]. It is the fastest algorithm among all the other algorithms [9]. It doesn't require any extra space like other sorting algorithms. For this reason, it has a variety of applications. Quick sort [16] uses the divide-and-conquer technique. It requires a pivot element to be selected from the input numbers. The array is then divided into sub-arrays. This process is called partitioning. Quick sort reconstructs the original array with the left part having elements whose value is less than the pivot and the right part having elements whose value is greater than the pivot. The algorithm repeats this operation recursively for both the sub-arrays. Quick sort has the worst-case time complexity of $O(n^2)$ and the average-case, and best-case time complexity of $O(n\log(n))$.

The seventh algorithm is radix sort [17]. It is not a comparison-based algorithm like insertion sort and quick sort. It works by sorting each digit on the input element and for each of the digits in that element. The sorting starts from the digit at the one's position and continues till the entire length of the number. Radix sort is a stable sort. In all the cases, the time complexity of this algorithm is $O(kn)$. Table 13.1 presents the comparison of various sorting algorithms based on time and space complexity.

13.3   PROPOSED ALGORITHM

The proposed algorithm is given in the form of pseudocode and flowchart in Sections 3.1 and 3.2, respectively. It is implemented using the Java programming language. The algorithm can also be implemented in any other language. To implement the algorithm, positive integers are stored in a data structure (array) [19]. The entire array is traversed to find the largest element in it. After traversing the array, a new array is initialized, whose size is one greater than the maximum element. The reason is that the array starts with index 0. After initializing the array, the original array is

TABLE 13.1
Comparison of Different Sorting Algorithms

Sorting Algorithm	Time Complexity			Space Complexity	Stability
	Worst Case	Average Case	Best Case		
Quick Sort [18]	$O(n^2)$	$O(n\log n)$	$O(n\log n)$	$O(n\log n)$	No
Merge Sort [18]	$O(n\log n)$	$O(n\log n)$	$O(n\log n)$	$O(n)$	Yes
Bubble sort [18]	$O(n^2)$	$O(n^2)$	$O(n)$	$O(1)$	Yes
Insertion Sort [18]	$O(n^2)$	$O(n^2)$	$O(n)$	$O(1)$	Yes
Selection Sort [3]	$O(n^2)$	$O(n^2)$	$O(n^2)$	$O(1)$	No
Radix sort [18] k:range of elements	$O(kn)$	$O(kn)$	$O(kn)$	$O(n+k)$	No
Heap Sort [18]	$O(n\log n)$	$O(n\log n)$	$O(n\log n)$	$O(1)$	No

traversed where initial data was stored. It updates the frequency of the element of the old array in the newly instantiated array. For example, if the number is 4, then the value in the array at index 4 is incremented. The above steps are repeated for each element. After this process, it traverses through the new array. This time, if the value stored in the array is 0, then it continues to the next iteration; otherwise, it stores the index of the number in the original array depending upon the value in the current array. For example, if the value is 3 at index 5 of the array, then store 5 three times in the array. After completing this process, all the values in the array will be 0 and the original array will be sorted.

13.3.1 Pseudocode

```
Algorithm MyAlgorithm(arr,n)
    // Input: 'arr' is the list of numbers to be sorted, 'n' is
the size of array
    largest=arr[0] // Initializing the largest element as the
first element
    i=1
    While i is less than n:
        if arr[i] is greater than largest:
            largest=arr[i]
        i=i+1
    Initialize/Create a new 'store' array with size of largest+1
    i=0
    While i is less than n:
        store[arr[i]]=store[arr[i]]+1;
        i=i+1
    i=0,k=0
    While i is less than equal to largest:
        if store[i] is not equal to 0:
            temp=store[i]
            While temp is greater than 0:
                arr[k]=i
                 temp=temp-1;
                k=k+1
    i=0
    While i is less than n:  // Printing the array
        print(arr[i])
        i=i+1
```

13.3.2 Flowchart

The flowchart is depicted in Figure 13.1.

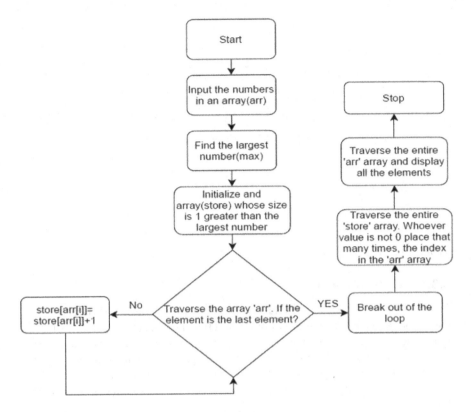

FIGURE 13.1 Flowchart of the proposed algorithm.

13.3.3 EXPLANATION OF PROPOSED ALGORITHM

Let's take an example. Suppose the input elements are {4,2,3,2,1,4,5,5,6,7}. First, store the input in an array. Then, find the largest element among these numbers. the largest element in the array is 7. Now, a new array "store" with size 8 is created. After this, again traverse the old one array arr[]. This time, update the value in the new array store []. In the first iteration, the value is 4 in the arr[] array. So, update the index position 4 in store [] array to 1. In the next iteration, value 2 is encountered in the arr[] array. So, the index position 2 in the store [] array is updated to 1. In the third iteration, update the index 3 in the store [] array to value 1. In the fourth iteration, update the index 2 of the store [] array to value 2, i.e., 1+1. Perform this for the rest of the elements as shown in Step 3 of Figure 13.2. In the final step, traverse the store [] array. Ignore all the values which have 0 in them. Therefore, index position 0 is ignored. In the next iteration, the value is 1 so store index position 1 the arr[] once. Then in the next iteration, the value found is 2. So, store index position 2 in the arr[] array twice. Then in the next iteration, value 1 is encountered. So, store 3 once in the arr[] array. This will continue till the entire array is sorted. Continuing this, which will lead to a sorted array (see Figure 13.2).

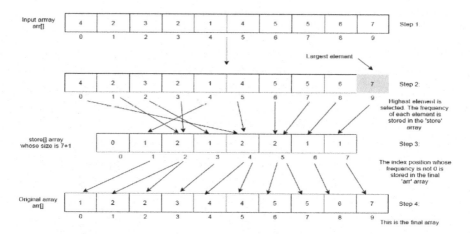

FIGURE 13.2   Explanation.

13.3.4 Algorithm Analysis

The algorithm has been tested under various conditions and for different sets of inputs. The time taken by our algorithm to sort the given list is compared with existing algorithms, and the obtained results are compared. Four different cases (best, average, worst, and sorted) are considered to compare the obtained results. In each case, the algorithm is subjected to different input sizes. The results obtained for these four cases are presented in Tables 13.2–13.5, which are also plotted as a graph (see Figures 13.3–13.6).

13.3.4.1   Best Case

For plotting the graph, a scale of 2000 is used on the *y*-axis (execution time in nanoseconds) and 50 on the *x*-axis (size of data set). To test the proposed algorithm for the best case, integers ranging from 1 to 10 were taken. Different input sizes were taken, and the execution time of the proposed algorithm is compared with the execution time of the existing algorithms. Repetition of the integers is allowed in the best case. The numbers present in the best case aren't sorted and are in completely random order. The proposed algorithm outperformed all the existing algorithms in this case.

TABLE 13.2
Best-Case Comparison with Existing Algorithms

Data Set Size	Bubble Sort	Heap Sort	Insertion Sort	Merge Sort	Proposed Algorithm	Quick Sort	Selection Sort	Radix Sort
10	4400	6800	3400	8800	2500	6200	4700	35100
100	135200	84200	52400	53400	6800	35800	90100	72400
1000	6264300	639800	3209000	1202100	69700	616600	4113500	375500
5000	49647800	1075400	12493300	2051100	194700	2844700	17744900	848700
10000	196189300	1742100	18852700	1849300	485800	15662700	71289400	3607500
50000	4861125700	6686400	294147800	10212400	1953500	53502700	1771793000	5410400
100000	19003803100	12666500	968398300	14225000	1478000	212042100	6534957700	10017700

TABLE 13.3
Average-Case Comparison with Existing Algorithms

Data Set Size	Bubble Sort	Heap Sort	Insertion Sort	Merge Sort	Proposed Algorithm	Quick Sort	Selection Sort	Radix Sort
10	3700	6200	2900	8600	12700	6200	3400	47600
100	146100	56000	50000	56800	18600	30100	76200	72700
1000	6171800	709800	3609500	681500	63600	713900	4272100	421000
5000	52640700	1137900	7727900	1343900	215700	1091500	24239100	2961500
10000	189377600	2255200	16311200	2567300	729800	1744200	69057200	3409700
50000	4761851400	7635200	335253900	9253400	1994400	8687000	1686451900	7798800
100000	18791540800	14711500	1015904900	14795800	904200	11544000	6453579500	15452500

TABLE 13.4
Worst-Case Comparison with Existing Algorithms

Data Set Size	Bubble Sort	Heap Sort	Insertion Sort	Merge Sort	Proposed Algorithm	Quick Sort	Selection Sort	Radix Sort
10	3900	19500	7700	7800	6278500	20600	4500	102800
100	143300	81900	111600	56100	7985500	33100	88200	83400
1000	13759100	777800	4462300	934800	7607900	1696300	4380900	551100
5000	49765300	1044100	11723400	1499600	6586400	1129100	13997400	3662600
10000	190393400	1858700	21538300	2469700	8355600	1600300	67915700	5496200
50000	4793619400	8357900	275580500	14679100	7821600	7874800	1683461600	12191400
100000	18923255700	15277600	1008332000	17971200	5036400	14820000	6435655800	50515900

Table 13.5
Sorted-Case Comparison with Existing Algorithms

Data Set Size	Bubble Sort	Heap Sort	Insertion Sort	Merge Sort	Proposed Algorithm	Quick Sort	Selection Sort	Radix Sort
10	3200	6200	2300	7700	1700	7800	3200	25500
100	89900	58300	4600	52200	6900	222200	73500	46900
1000	5120300	734400	26500	1060300	117500	8585300	608000	3998300
100000	24189600	1645400	253900	1890900	1118600	24943500	2829500	22389900

The execution time taken by the proposed algorithm, in this case, is much less than all existing algorithms. The space complexity of the proposed algorithm, in this case, is $O(K)$, where K is the maximum element in the array. Since all the numbers are from 1 to 10, in the worst case, the space complexity is $O(11)$ irrespective of the input size.

13.3.4.2   Average Case
For plotting the graph shown in Figure 13.4, the scale on the y-axis is taken as 2000 and the scale on the x-axis is 50. For testing the proposed algorithm, integers ranging from 1 to 10000 are taken. The numbers are in random order, and repetition of the numbers is also allowed. The algorithm is tested for a different size of the

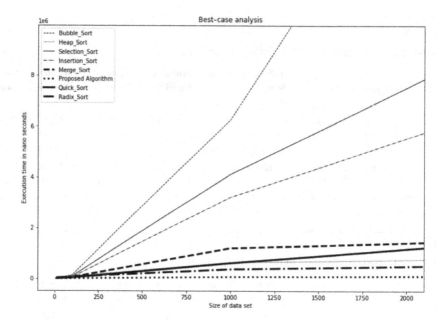

FIGURE 13.3   Best-case analysis of the proposed algorithm.

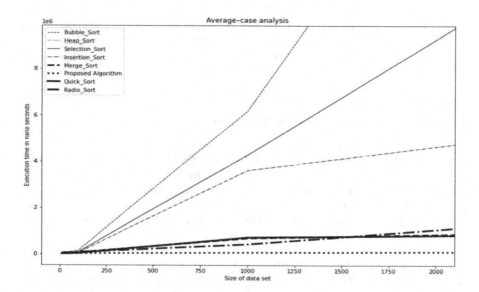

FIGURE 13.4   Average-case analysis of the proposed algorithm.

input. When the input size is less, the execution time of the proposed algorithm is bad as compared to the execution time of other algorithms. But when the input size becomes greater than 100, the algorithm starts performing better than all the existing algorithms. In the worst case, an array with size 10001 is required. This will happen

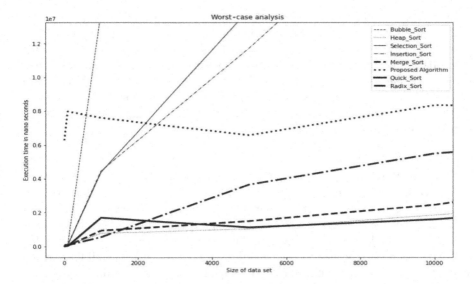

FIGURE 13.5 Worst-case analysis of the proposed algorithm.

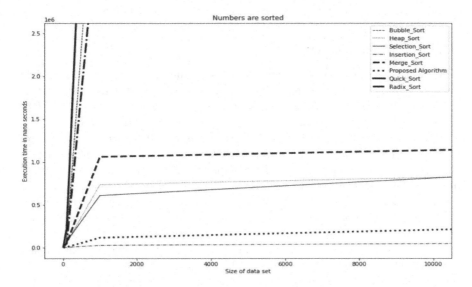

FIGURE 13.6 Sorted-case analysis of the proposed algorithm.

when there is at least one element whose value is 10000. Otherwise, the array size will be one more than the highest element.

13.3.4.3 Worst Case

The graph shown in Figure 13.5 has been plotted with a scale of 1500 on the y-axis and 10 on the x-axis. A data set in which all the numbers are integers ranging from 1 to 1000000 is taken to test the algorithm. Repetition of the numbers is allowed. The

numbers are in random order and aren't sorted. The execution time of the proposed
algorithm is compared with the execution time of all the existing algorithms. The
proposed algorithm's performance is worst for small input sizes. But as the input
size increased around 10000, the proposed algorithm started performing better than
some existing algorithms. In the worst case, the execution time is bad. The extra
space taken during this time is also around 1000001 even for a small size input. From
this, it can be concluded that, when the numbers are large, the algorithm won't be
efficient for sorting the numbers.

13.3.4.4   Sorted Case

A scale of 10 has been used on both axes to plot the numbers (see Figure 13.6). A
data set in which numbers are from 1 to 10000 is taken to test the algorithm. The
numbers are sorted in ascending order. No repetition of the numbers is allowed in
this case, i.e., the numbers are distinct. The running time of the proposed algorithm
is compared with all the existing algorithms. From the obtained results (see Table
13.5), we can say that the proposed algorithm performs better than all the existing
algorithms except insertion sort. Even if the numbers are sorted or they are distinct,
the proposed algorithm performs better. The size of the additional array in the sorted
case depends on the maximum number present in the input. In the worst case, an
additional array of size 10001 irrespective of the input size is required.

13.3.5   EXPERIMENTAL EVALUATION

The experiment (execution of the proposed and existing algorithms) is conducted on
an Intel Core i5 processor with a 2.4 GHz processor. The RAM of the system was 8
GB with Windows 10 as the only operating system. The code is written in Java language, and Visual Studio Code is used as the text editor for testing purposes.

As mentioned in Tables 13.2–13.5, the execution time of the proposed algorithm
depends on the maximum element from the given input. If the maximum element is
very large and the input size is small, then the algorithm is not efficient. But when
the maximum element is small, then irrespective of the input size the algorithm is
better than most of the existing algorithms and in some cases all of them. The proposed algorithm is not a stable sorting algorithm as the order of insertion of elements
is not kept in the account. This algorithm can be made a stable one but then its time
complexity will increase. The algorithm performs well even when the numbers are
repeated. The only drawback of the algorithm is that it requires extra space for storing the frequency of the element. That extra space is dependent upon the highest
element. If the highest element in the input is less, then the extra space consumed
will be very less.

13.3.6   LIMITATIONS

In the proposed algorithm a hash-based approach is used. The frequency of the element is stored in an additional array. So, there are certain limitations in the algorithm. This would result in undesirable outputs. The following are the limitations of
the proposed algorithm:

1. The algorithm won't work with negative integers. A hash-based approach is followed in the algorithm, which uses an array to store numbers. An array doesn't have negative indices to store the frequency of negative elements. The algorithm can be modified to perform sorting of negative numbers as well. A possible solution is mentioned in the future scope.
2. The algorithm only works for integers. The algorithm can't sort floating-point numbers as an array has been used to store the frequency. Array doesn't support the fractional value of indices.
3. The algorithm can only sort numbers, which can be stored in an 'int' data type of any language. To make it flexible for larger size, certain modifications must be made.
4. The algorithm is not a stable sort like bubble, merge, and insertion sort.

13.3.7 FUTURE SCOPE

Some modifications can be made in the proposed algorithm to improve its time as well as space complexity and also to adapt various types of numbers. The following modifications are suggested:

1. To make the algorithm work for a negative number, the frequency of numbers should be stored in a 'HashMap' [20] of Java language or any other similar data structure in any other language. Then, the map must be sorted based on the key. After sorting, the elements must be stored back in the original array. This will increase the time a little bit. But if the sorting is done efficiently, the time won't be increase dramatically.
2. The proposed algorithm can be made stable by using an array of queues. The element can be added to the queue of the array. The queue data structure will preserve the order of insertion. Hence the algorithm will become stable. However, this will increase the space as well as the time complexity of the algorithm.
3. 'TreeMap' [21] in Java can be used to store the frequency of the elements. This will save space, but an insertion in a 'TreeMap' will take order of $\log(n)$ time.

13.4 CONCLUSION

Thus, this paper realizes a new sorting algorithm based on hashing approach. After analyzing and testing, the proposed algorithm performs better than most of the other algorithms in its best-case scenario, i.e., when the input size is nearly equal to the maximum element. But, in its worst-case scenario, when the largest element is much greater than the input size, its performance decreases compared with radix sort, heap sort, quick sort, and merge sort algorithms.

REFERENCES

1. Comparison-Based Sorting, https://opendatastructures.org/ods-java/11_1_Comparison_Based_Sorti.html, last accessed on 2021/10/13.
2. What Does 'Space Complexity' Mean? – GeeksforGeeks, https://www.geeksforgeeks.org/g-fact-86/, last accessed on 2021/09/02.
3. Stability in Sorting Algorithms – GeeksforGeeks, https://www.geeksforgeeks.org/stability-in-sorting-algorithms/, last accessed on 2021/09/09.
4. Sedgewick, R., & Flajolet, P. (2013). *An Introduction to the Analysis of Algorithms.* Pearson Education India. https://dl.acm.org/doi/abs/10.5555/227351.
5. S. Bae. (2019). Big-O Notation: An Introduction to Understanding and Implementing Core Data Structure and Algorithm Fundamentals. Doi: 10.1007/978-1-4842-3988-9_1.
6. A. Karunanithi, and F. Drewes. A Survey, Discussion and Comparison of Sorting Algorithms, Umea University, June 2014.
7. Zutshi, A., & Goswami, D. (2021). Systematic review and exploration of new avenues for sorting algorithm. *International Journal of Information Management Data Insights*, 1(2), 100042. https://www.sciencedirect.com/science/article/pii/S2667096821000355.
8. Bubble Sort – GeeksforGeeks, https://www.geeksforgeeks.org/bubble-sort/, Insertion Sort – GeeksforGeeks, last accessed on 2021/09/03.
9. Y. Yang, P. Yu, and Y. Gan. "Experimental study on the five sort algorithms," *2011 Second International Conference on Mechanic Automation and Control Engineering*, 2011, pp. 1314–1317, doi: 10.1109/MACE.2011.5987184.
10. Insertion Sort - GeeksforGeeks, https://www.geeksforgeeks.org/insertion-sort/, last accessed on 2021/09/03.
11. S., Jadoon, S. F. Solehria, and M. Qayyum. (2011). Optimized selection sort algorithm is faster than insertion sort algorithm: A comparative study. *International Journal of Electrical & Computer Sciences*, 11, 19–24.
12. Selection Sort – GeeksforGeeks, https://www.geeksforgeeks.org/selection-sort/, last accessed on 2021/09/03.
13. HeapSort - GeeksforGeeks, https://www.geeksforgeeks.org/heap-sort/, last accessed on 2021/09/03.
14. Merge Sort - GeeksforGeeks, https://www.geeksforgeeks.org/merge-sort/, last accessed on 2021/09/03.
15. QuickSort - GeeksforGeeks, https://www.geeksforgeeks.org/quick-sort/, last accessed on 2021/09/03.
16. C. A. R. Hoare. (1962). Quicksort. *The Computer Journal*, 5(1), 10–15.
17. Radix Sort - GeeksforGeeks, https://www.geeksforgeeks.org/radix-sort/, last accessed on 2021/09/03.
18. P. Prajapati, N. Bhatt, and N. Bhatt. (2017). Performance comparison of different sorting algorithms. *International Journal of Latest Technology in Engineering, Management & Applied Science*, VI(Vi), 39–41.
19. Array Data Structure - GeeksforGeeks, https://www.geeksforgeeks.org/array-data-structure/, last accessed on 2021/09/02.
20. HashMap in Java with Examples – GeeksforGeeks, https://www.geeksforgeeks.org/java-util-hashmap-in-java-with-examples/, last accessed on 2021/09/07.
21. TreeMap in Java – javatpoint, https://www.javatpoint.com/java-treemap, last accessed on 2021/09/07.

14 A New Proposed Sorting Algorithm Using Radix Tree and Binning

Kashvi Dedhia, Riddham Gadia, Saloni Dagli, Shivam Kejriwal, and Stevina Correia
Dwarkadas J. Sanghvi College of Engineering

CONTENTS

14.1 INTRODUCTION

With an increase in the methods of generation of data came an increase in the means of sorting this generated data. This gave rise to a new field of study called sorting. These algorithms have been developed over long periods and have required the researchers to put an immense amount of effort and time into developing and testing the algorithms. These implemented algorithms are used by the layman in their daily lives. An example of this is the sort by that is provided in the e-commerce portals. Herein, the portals make use of a combination of sorting algorithms in order to sort the data in an efficient way. Another example from our daily lives that makes the use of sorting is the ranking systems that are used in schools and colleges. These systems aggregate the marks of all the students and sort them in ascending order after which they assign the students their respective ranks. Some ways in which we can determine the performance of sorting algorithms are by analyzing them with respect to the time taken and space occupied. We try to see how much time a particular sorting algorithm takes, based on the number of digits that need to be sorted.

DOI: 10.1201/9781003320333-14

The time complexity of a database can vary widely, even for the same number of digits. Therefore, it is best to check the time complexity in the best case, average case, and worst case. However, even after calculating the time complexity, we need to test the algorithm to see how much time the algorithm takes. Similarly, the space complexity is tested and estimated, that is, the amount of space taken by the algorithm while running, with respect to the input size. The algorithm proposed in this paper uses a radix tree to sort numbers initially, followed by using bins to finally get the sorted output. This algorithm and other common sorting algorithms are tested with respect to the time, for datasets of varying size and distribution.

14.2   LITERATURE REVIEW

Selection sort is a type of sorting in which at each iteration, the minimum element is selected, and it is inserted at the appropriate location. This algorithm divides the array into two parts. The first is the sorted array and the second is the unsorted array. The minimum element is selected in the unsorted array, and then it is placed at the appropriate position in the sorted array [1]. This algorithm continues till the entire list is sorted. It is a comparison-based algorithm. The running time complexity of this algorithm is $O(n^2)$ [1,2]. In the worst case, the algorithm will need $n(n-1)/2$ [3] comparisons. The space complexity of the algorithm is $O(1)$. The disadvantage of the algorithm is that it will in the best case also it will have a time complexity of around $O(n^2)$. There are various improvements in the running time of selection sort [4,5] but still, in the worst case, it performs badly for a large set of data [3,6].

Radix sort is a type of sorting algorithm in which numbers are sorted based upon the position. The least significant digit is sorted first, and then it continues sorting the next least significant digit until the most significant digit is reached [7,8]. Radix sort involves the use of buckets to store the data. Radix sort can be used for both sorting numbers as well as sorting strings. In the case of numbers, the base will be 10 since there are 10 digits (0–9), and in the case of strings, the base will be 26 since there are 26 alphabets (A-Z). The running time of radix sort is dependent on the number of digits in the input number [9]. The running time complexity of radix sort is $O(n+k)$, where 'n' stands for the size of the integer dataset and 'k' is the maximum digit size for a given set of numbers or strings [8]. Radix sort is not dependent on the size of the data, but it is dependent on the number of digits in the maximum element [10]. This is one of the main disadvantages of Radix sort as even if the number of data is less, if the number of digits in the maximum element is huge, its execution time will increase [6].

Shell sort is a modification of insertion sort. It is highly suitable for medium-sized datasets [11]. The time complexity of shell sort is $O(n)$ for the best as well as the worst case scenario. It starts with sorting all the elements that are a certain length apart from one another, also called the interval. The interval is then divided by 2, and the same process continues, till the interval reaches 1, after which it stops. The first interval is usually taken as half of the size of the dataset. In each round, the position of the number which is selected is saved and then compared to the number that is interval length away. If it is lesser than the other number, it stays in the same position. However, if the number is greater than the other number, then it exchanges places. After this, the original number is compared to the number which is now interval length away to it. The same comparison process continues till a number is found which is greater than it or there is no number left that is interval length away. After this, the number selected changes to the one at the next position from the one that is saved. The same process continues with all the numbers for the same interval. The whole process is repeated from the first to the last number, for every interval, till the interval becomes 1. The output after this is a sorted list.

Insertion sort [12] is an in-place sorting algorithm. The array of elements to be sorted is virtually divided into two sections – sorted and unsorted. Elements from the unsorted section are moved and placed at the correct position in the sorted section one by one. Consequently, in each iteration, the size of the sorted array keeps increasing. It is relatively less efficient on large arrays than most advanced algorithms available today. Hence, it works best with a smaller number of elements or in cases where only a few elements need to be sorted from a massive array. Its worst case time complexity is $O(n^2)$.

Bubble sort is a simple algorithm to sort an array of n elements by their values. It is a type of representation that shows how the biggest element in an input array rises to the top index after each complete iteration [13]. For example, if an array needs to be arranged in ascending order, bubble sort would start by comparing the first two elements, swapping the two if the first element is greater than the second.

A radix tree is a type of trie data structure wherein every single child is merged with its parent node. The property that differentiates a radix tree from a normal trie is that it uses both multiple elements as well as single elements structures for labeling the edges. This property results in the keys of the tree being compared in sets rather than individually which in turn helps in reducing the amount of time required for searching. Insertion in a radix tree involves searching the tree chunk by chunk to check whether the chunks correspond to the string. As soon as there is a mismatch, we terminate the searching operation and insert the rest of the elements at a new edge node. In case there is already an edge-sharing the same value of the prefix as that of the string, then we split the prefix and insert the rest of the suffix. Deletion is relatively simpler. In order to delete a string value from a tree, we first find the string in the tree. Once we find the string, we delete the leaf node and, in the case, where the parent of that node has only one child leaving aside the leaf node, the label of the child is appended to that of the parent after which the child is removed.

14.3 PROPOSED ALGORITHM

The proposed algorithm has been explained with the help of flow charts. Figure 14.1 shows the flow of the proposed algorithm While Figure 14.2, in more detail shows how the numbers are inserted into the radix tree. In our proposed algorithm, first, the numbers are input as strings. They are then inserted into the radix tree. The radix tree is then traversed, and the output from the radix tree is saved into an array, arr[] as shown in Figure 14.1.

As shown in Figure 14.2, all the input numbers are stored in the radix tree. When the first element is inserted into the radix tree, the element is stored as it is. Then, for the second element, a mismatched character is found. If there are no mismatched

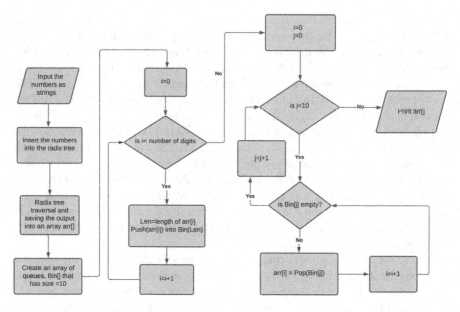

FIGURE 14.1   Flowchart of the proposed algorithm.

characters, the number is stored as it is. If a mismatched character is found, the root is split at that character and the remaining part of the number is stored as a child of that node. The isLeaf flag is set to true at the child. All the numbers are inserted into the radix tree in this manner. Then, a preorder traversal of the radix tree is performed. Whenever the isLeaf flag is set, the numbers are stored in an array in the same order. This will give us a partially sorted output.

This output is partially sorted; however, it is not fully sorted. To sort the output fully, we need to perform binning, as shown in Figure 14.1. A bin is an array of 10 queues as shown in Figure 14.1. For every number of the dataset, the number of digits or the length of the string is found. Based on the length, it is pushed into the corresponding queue. If the length of the string is 2 of the number has two digits, then it is pushed into the second queue or Bin 2. This continues for every number in the array arr[]. After this, the bins are traversed. First, all the numbers in the first queue or Bin 1 are popped. Therefore, all the one-digit numbers are printed, in the same order as they appeared in the output from the traversal of the radix tree. Then, after it becomes empty, the same is done for the next queue. This process repeats for all the queues from Bin 1 to Bin 10 in increasing order, till they are all empty. As and when a number is popped from the queue, it is saved in an array. The output of the array is the final sorted answer. The bin has 10 queues because we have taken numbers that have a maximum of 10 digits, in our dataset. Depending on the range of the dataset, the number of queues can be adjusted.

For example, in Figure 14.3, we take a dataset with 13 numbers that need to be sorted. Below shows the first step after inputting the numbers as strings. It shows the insertion of the numbers into the radix tree. One by one the numbers are inserted

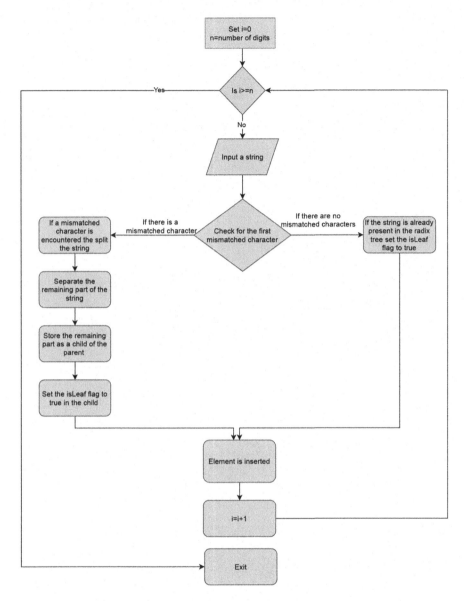

FIGURE 14.2 Flowchart of insertion of the numbers into the radix tree.

into the radix tree, as every step shows. After the radix tree is completely formed, it is traversed. The output from the traversal is saved into an array.

After this, every step shows the insertion of the dataset into the respective queues of the Bin as shown in Figure 14.4. After the insertion is complete, the path taken by the traversal is shown by the green arrow as shown in Figure 14.5. The output is a sorted array.

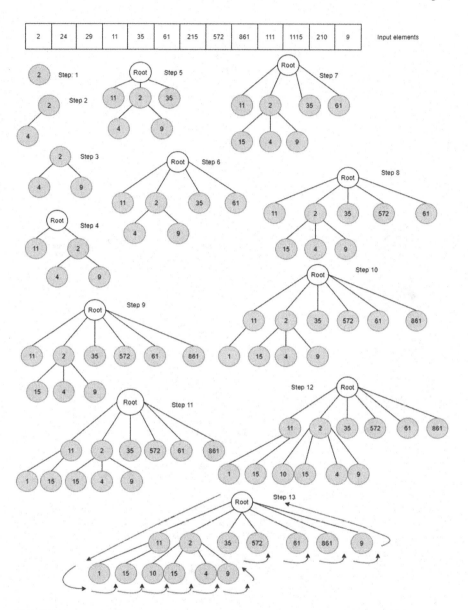

FIGURE 14.3 Illustration showing how the numbers would be inserted into and traversed from the radix tree using the example dataset.

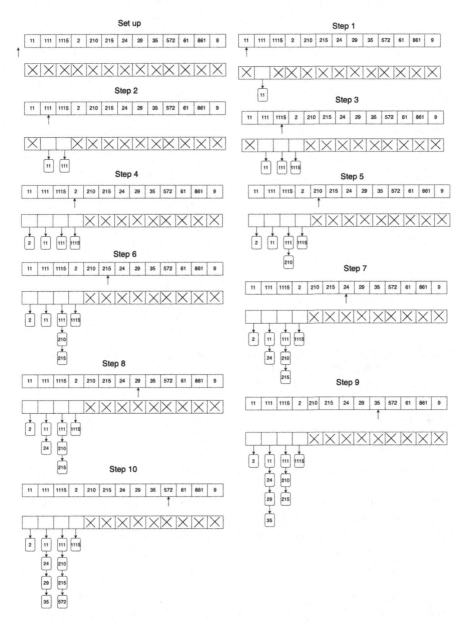

FIGURE 14.4   Illustration showing how the numbers would be inserted into the bins, after the output is obtained from the radix tree, till step 9.

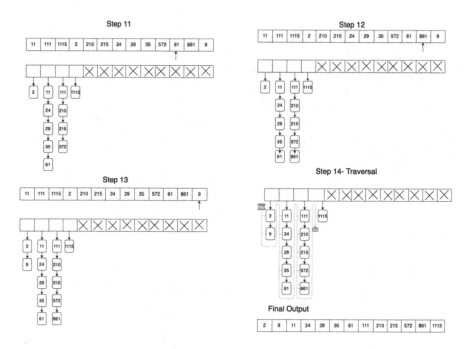

FIGURE 14.5 Illustration showing how the numbers would be inserted from step 10 to step 13, along with showing the traversal of the bin and final output.

14.4 DATASET

14.4.1 TYPE 1 DATASET

The proposed algorithm has been tested for various input sizes. In this dataset, the numbers were from the range 1–10000000. All the numbers were integers and were distinct. The numbers are in random order. To test whether the proposed algorithm was dependent on the order of inputs, multiple cases were taken for each input size. Some of the datasets were sorted, and some were in random order.

14.4.2 TYPE 2 DATASET

The proposed algorithm works differently for a particular type of input. So, in this dataset, the proposed algorithm was tested taking very few digits and testing various permutations of it. For example, digits 1,2,3,4 were taken and all the permutations of it were taken into account. The length of the number could be anything, but it contained only four digits in this case. In this way, the algorithm was tested for numbers when there is a pattern. The number of digits was changed each time for different input sizes. The numbers were distinct. For example, if we take digits 2 and 3. So the various permutations of the digits are 2,3,22,33,23,32,223,323,232,333,222,2223, and so on.

14.5 EXPERIMENTAL SETUP

The above experiments were conducted in an Intel Core i5 processor with a 2.4 GHz processor. The RAM of the system was 8 GB. It was conducted on a laptop with Windows 10 as the only operating system. The code was written in Java language, and Visual Studio Code was used as the text editor for testing purposes.

14.6 PERFORMANCE EVALUATION

Table 14.1 shows the output obtained for testing with the type 1 datasets. The proposed algorithm is tested, along with bubble sort, insertion sort, radix sort, selection sort, and shell sort, with respect to the time taken to sort the dataset. The time has been calculated in nanoseconds. We have tested the algorithms with 15 sizes of the dataset from 10 to 50,000.

Comparison has been made with bubble sort, selection sort, and insertion as both the algorithm has a time complexity of $O(n^2)$. Shell sort has a time complexity of O(nlogn) in the average case and radix sort has a time complexity of $O(d*(n+b))$, where d is the number of digits in the given list, n is the number of elements in the list, and b is the base or bucket size used. This ensures an unprejudiced comparison among all the elements.

Figures 14.6 and 14.7 shows the output for the proposed algorithm, along with other commonly used algorithms. The graph in Figure 14.7 is made by using the

TABLE 14.1
The Number of Nanoseconds that All Six Algorithms Take to Sort Type 1 Datasets

Test Case Size	Our Algorithm	Bubble Sort	Insertion Sort	Radix Sort	Selection Sort	Shell Sort
10	1081300	4400	3800	42800	6200	7000
10	1162600	4100	4200	47400	5900	6800
20	1268500	8200	5200	49400	17300	15500
20	1227000	8100	5500	47100	13300	12000
100	1626500	125200	43500	154200	193200	45800
100	1612300	127700	48200	92400	179900	33700
200	2161100	589900	183800	281400	557700	68000
200	2130800	509900	184200	314700	537300	65100
500	3527400	2026700	1028300	679700	1747100	184100
500	3402200	1970200	1019300	516500	1544700	184000
1000	5978100	7082100	4730000	821400	4471400	611100
1000	6405100	7261900	4614600	798500	5882500	541900
10000	12853900	136324200	99573400	6122700	12721700	7748000
10000	11094400	132092100	92328300	6088000	12901800	8100200
50000	74777900	320901300	223000900	39845000	130719200	10444000

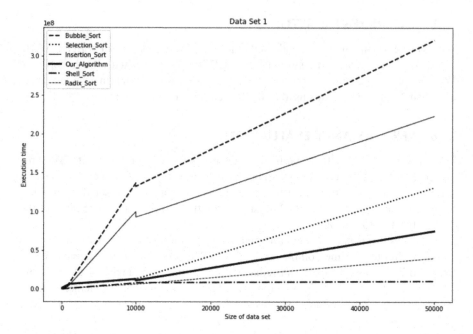

FIGURE 14.6 Time taken to sort vs the size of the dataset for type 1 dataset for all six algorithms.

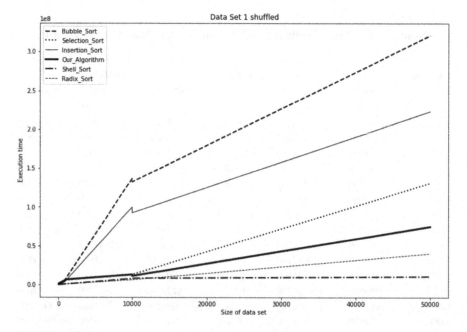

FIGURE 14.7 Time taken to sort versus the size of the dataset for type 1 shuffled dataset for all six algorithms.

TABLE 14.2

The Number of Nanoseconds that All Six Algorithms Take to Sort Type 2 Datasets

Test Case Size	Our Algorithm	Bubble Sort	Insertion Sort	Radix Sort	Selection Sort	Shell Sort
10	1022000	4300	6800	46400	6100	6000
10	1182700	5200	3900	36200	5800	6200
20	1309900	9500	5900	49500	11700	9900
20	1235700	8800	4980	41000	11000	11100
100	177000	159000	50400	85100	228400	32200
100	205000	159100	47500	81900	199700	32700
100	190800	134900	69400	104200	195100	28600
500	3510000	1841200	1122500	336900	1662300	176700
500	3324500	1754800	1743400	315900	1445600	174800
1000	5997000	5764800	4969100	583000	4887900	461500
1000	6592600	6004200	4825400	498900	3998800	432700
8429	29822900	78606700	63475400	3672400	30499100	3261900
9507	28908000	102605300	97943300	3304500	34789200	3655800
50000	55472800	380203100	178982300	21074000	120480400	10249200

same dataset as Figure 14.6; however, the numbers have been shuffled. The output for both graphs is about the same. The proposed algorithm performs better than bubble sort, insertion sort, and selection sort in all cases where the number of numbers were more than 1000. The slope is increasing; however, it is doing so very slowly. Therefore, this algorithm will be useful in cases where n is a large number.

Table 14.2 shows the output obtained for testing with the type 2 datasets. The proposed algorithm is tested, along with bubble sort, insertion sort, radix sort, selection sort, and shell sort, with respect to the time taken to sort the dataset. The time has been calculated in nanoseconds. We have tested the algorithms with 15 sizes of the dataset from 10 to 50,000.

Figures 14.8 and 14.9 show the output for the proposed algorithm, along with other commonly used algorithms. The graph in Figure 14.9 is made by using the same dataset as the graph in Figure 8; however, the numbers have been shuffled. The output for both graphs is about the same. The proposed algorithm performs better than bubble sort, insertion sort and selection sort in all cases where the number of numbers were more than 10,000. The slope is increasing; however, it is doing so extremely slowly, almost constant. This can be very useful for large datasets.

14.7 LIMITATIONS

The limitations of the proposed algorithms are as follows:

- The proposed algorithm won't work with negative numbers. To implement the algorithm, a radix tree has been used. Radix tree doesn't support negative numbers.

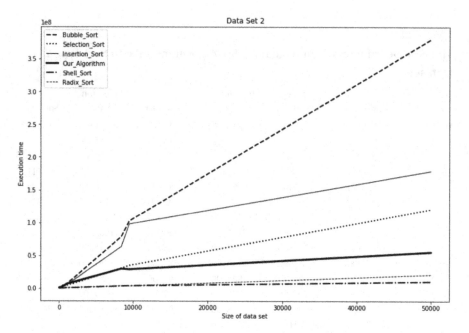

FIGURE 14.8 Time taken to sort versus the size of the dataset for type 2 dataset for all six algorithms.

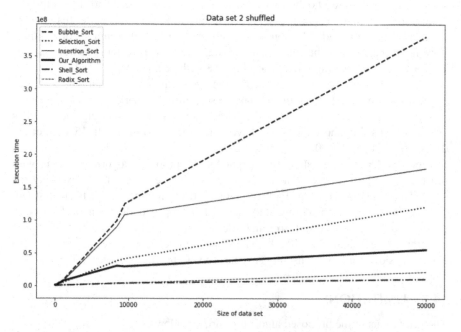

FIGURE 14.9 Time taken to sort versus the size of the dataset for type 1 shuffled dataset for all six algorithms.

- The algorithm won't be efficient when the input size is small. The algorithm will only be efficient when there are many elements to be sorted.
- The algorithm won't work when there is a repetition of numbers.
- The algorithm makes use of a radix tree. The algorithm won't be efficient when the space is very less.
- The algorithm can only work when the maximum length of the digits is 9. Beyond that, the algorithm will fail.

14.8   CONCLUSION

After performing the analysis and plotting the graphs, we can infer that the above proposed algorithm performs better than bubble sort, selection sort, and insertion sort in all cases wherein the size of the dataset is greater than 10000. Keeping in mind the limitations of the proposed algorithm, the results achieved were considerably good.

14.9   FUTURE SCOPE

Some modifications can be made to the algorithm to make it more efficient.

- A more optimized version of the radix tree can reduce the space as well as the time complexity of the algorithm.
- The algorithm can sort numbers greater than 1000000000 if a data structure like HashMap is used to store the number of digits.
- The algorithm can be modified to sort negative numbers by separating the negative numbers from the input and then again combing with the sorted positive integers.

REFERENCES

1. Ravendra Kumar: Review and Analysis of Sorting Techniques in Various Cases, *International Journal of Advanced Research in Computer Science and Software Engineering*, 652–655 (2016).
2. Ramin Edjlal; Armin Edjlal; Tayebeh Moradi: In: *2011 3rd International Conference on Computer Research and Development*, pp. 380–381, IEEE, Shanghai, China (2011).
3. You Yang; Ping Yu; Yan Gan: In: *2011 Second International Conference on Mechanic Automation and Control Engineering*, pp. 1314–1317, IEEE, Inner Mongolia, China (2011).
4. Sultanullah Jadoon, Salman Faiz Solehria, Salim ur Rehman, Hamid Jan: Design and Analysis of Optimized Selection Sort Algorithm, *IJECS: International Journal of Electrical and Computer Sciences*, 16–21 (2011).
5. Sultan Ullah; Muhammad A. Khan; Mudasser A. Khan; Habib Akbar; Syed S. Hassan: In: *2015 12th International Conference on Fuzzy Systems and Knowledge Discovery (FSKD)*, pp. 2549–2553, IEEE, Zhangjiajie, China (2015).
6. Purvi Prajapati, Nikita Bhatt, and Nirav Bhatt: Performance Comparison of Different Sorting Algorithms. *International Journal of Latest Technology in Engineering, Management & Applied Science (IJLTEMAS)*, 39–41 (2017).

7. Radix Sort Homepage, https://www.geeksforgeeks.org/radix-sort, last accessed 2021/09/09.

8. S. Anthony Vinay Kumar; Arti Arya: In: *2016 11th International Conference on Computer Engineering & Systems (ICCES)*, pp. 305–312, IEEE, Cairo, Egypt (2016).

9. Paul K. Mandal; Abhishek Verma: In: *2019 IEEE 10th Annual Ubiquitous Computing, Electronics & Mobile Communication Conference (UEMCON)*, pp. 149–153, IEEE, New York, NY, USA (2019).

10. Arman Bernard G. Santos; Melvin F. Ballera; Marmelo V. Abante; Neil P. Balba; Corazon B. Rebong; Bryan G. Dadiz. In: *2021 International Conference on Intelligent Technologies (CONIT)*, pp 1–6, IEEE, Hubli, India (2021).

11. Robert T. Smythe; Jon A. Wellner: Asymptotic Analysis of (3, 2, 1)-Shell Sort, *Random Structures and Algorithms*, 59–75 (2002).

12. Insertion Sort Homepage, https://www.geeksforgeeks.org/insertion-sort, last accessed 2021/09/04.

13. Bubble Sort Homepage, https://www.geeksforgeeks.org/bubble-sort, last accessed 2021/09/04.

15 Grey Wolf Optimizer for Load Balancing in Cloud Computing

Muskan
National Institute of Technology Hamirpur

Dharmendra Prasad Mahato
National Institute of Technology Hamirpur
Ton Duc Thang University

Van Huy Pham
Ton Duc Thang University

CONTENTS

15.1 INTRODUCTION

One of the most researched and talked about topic in network services over the past years is cloud computing. Resource pool provided which is available on the server and available by the other servers on requirement. Client can buy the cloud applications that the required by paying to how much time they used that particular application. As we know that the hike in the number of cloud clients over the past decade as well as bound on the number of required applications can cause the irregular distribution of users' requirement and also cause the other users to be overload and some of the application does not work without payment, which decrease the exertion of the application. Proper allocation of the services or user activities within all the users is required for balancing the load over the servers, so that no service provider or user will be overloaded and underloaded. The proper management of buyer requirements and tasks providers across every servers, so that we can balance the load properly over the internet. The tasks will be erased from the server if extra load is placed on it, considering the requirements of the virtual machines (VMs) and putting it on to the appropriate VM is one of the methods for assigning the task to the virtual machine. Based on the least distance between the server and client, it is allocated. There are so many algorithms that have been proposed to distribute the burden on the cloud environment, and there are so many research continues to improve these existing methods of load balancing in cloud computing. Containerization is the one the latest work in cloud computing that has better response as compared to other classical ways. Container is required to add the application such as Operating System (OS) levelization, but the virtual machine supports hardware virtualization. So to start a container, time required is much faster (per second) as compared to the virtual machine (per minute). Container is simple and weighty. It works better as compared to the virtual machines and the use of application with container is easier than the virtual machine. In the container cloud services, the aim of selecting a container from the file a collection of containers available where the work will be is done by looking at the load of all other containers. In this project, we use grey wolf optimizer (GWO)-based algorithm [1] for the distribution of uploads to a cloud containing container and reducing makespan. Utilization of the GWO algorithm and makespan. In this search, we will compare the load variation of our method to that of the genetic algorithm and the particle algorithm based on swarm optimization (PSO) (Figure 15.1).

15.2 PROBLEM STATEMENT

Load balancing in every environment required optimizing the throughput, resource/application uses, load imbalance, response time, overutilization, and underutilization of services. Due to increase in requirements of resources, providing proper service is one of the most challenging thing to do in cloud, since balancing the load on the server is the NP-Hard problem in heterogeneous circumstances.

15.3 OBJECTIVES

- To study and comparing the performance utilization of our algorithm with some of the existing load balancing algorithms, that is, genetic and PSO.

FIGURE 15.1 Cloud computing.

- Designing and development of the concepts of load balancing for the different sizes of the cloud.
- Analysing the status of virtual machines based on the calculated load which is balanced according to the GWO.
- Removal of task from the machine if extra load is placed on the virtual machine, and it will be assigned to the next best fit virtual machine, which is under loaded according to the assigned condition. And assignment of tasks to the virtual machine is purely based on the distance between the servers and nodes. If the distance is less, it is assigned to it.

15.4 BACKGROUND STUDIES

15.4.1 CLOUD COMPUTING

We know that cloud computing is one of the fastest evolving telematic over the internet which is mostly used with inside the network. As real-time commercial enterprise and IoT gadgets demand increasing day by day, so the organisations that want to increase and diminishing their offerings have started coming toward the cloud computing services providers on call for their services. As in a cloud services, the excessive extent of nonstop incoming demands for the services makes an imbalance in the cloud to records the load and provide smooth service. So current work is stabilising the weight over different server via optimizing the algorithm for deciding the most appropriate host and achieves balance of load over the server [2–9]. As nowadays demands of these services increasing day by day and providing these services properly over different server is one of the tedious job, balancing the load is must for providing better services [10–15].

- **Distribution of Nodes (Geographically):** As in the cloud environment, data centres are distributed geographically over the cloud for full filing the computational needs. In these, the nodes over the environment are considered as a single system which efficiently executes the user demands. Some of the load balancing algorithms are made for only small regions in which they do not take some factors under consideration such as networking problem, comm. problem between nodes and server, gap between the nodes across the network, gap between client and resources, and so on [16–21]. Nodes present at different regions is the problem, as some of these algorithms do not performed well for these type of behaviour. Thus, developing and implementing the load balancing algorithms for the nodes that are placed at different location and optimizing the utilization should be taken into consideration.
- **Failure of Nodes:** There are so many load balancing algorithms that are designed over past decades out of which some of them required non distribution and load balancing is decided by only the central node. If the central node does not work, then the whole computing environment is going to be crashed. So designing and developing some distributed algorithms for balancing the load in which only one node does not manage the whole computational process is required [22–27].
- **Migration of VMs:** Virtualization enables us to create several VMs on a single system. These VMs are completely independent of one another and have separate setups. If that single machine becomes overloaded, some of these Virtual Machines must be transferred to a new and more appropriate location via VM migration.
- **Heterogeneous Behaviour of Nodes:** As in the earlier stages of the cloud load balancing, only homogeneous nodes are taken into consideration. But now with evolving cloud computing, user demands changes dynamically which requires the execution of heterogeneous nodes for better utilization of resources and minimizing the response time as much as possible [28–36]. As a result, building an efficient load balancing method for a diverse environment is difficult.
- **Storage Management:** Cloud storage management eliminates the need for expensive hardware for personal storage, which was a concern with earlier conventional storage solutions. Now, the cloud allows all users to store their data heterogeneously throughout the network without any access issues [37]. However, cloud storage is growing in popularity, and keeping a replication of data over the internet for efficient access and consistency of data is becoming increasingly important. Full data replication techniques, as we know, are inefficient owing to data storage policy duplication on replication sites. While partial replication can suffice, dataset availability is a barrier since it raises the complexity of load balancing strategies. As a result, an effective load balancing methodology based on a partial replication system is necessary to be created in order to distribute applications and related data throughout the cloud.

- **Scalability of Balancer:** Because cloud services provide on-demand scalability and availability, customers may scale up and scale down services at any moment. And load balancer is considered good if it quick changes with the demands, that is, different sizes of the tasks, computational power, storage management and topology, and so on, and performed well in every environment.
- **Complexity of Algorithms:** Algorithms in cloud computing must be simple to build and comprehend. As the complexity of the algorithm increases, so will the systems performance and efficiency. As a result, developing a basic and straightforward algorithm is necessary.

15.4.2 LOAD BALANCING

Load balancing is utilised in many scheduling applications for proper job assignment to the server, such as in parallel, dispersed, and networked scenarios [38]. Because load balancing in store scheduling has not yet been thoroughly examined. The goal is to lower the pressure on computers in this manner so that we may increase gadget utilisation while also minimising total job throughput time. This may be accomplished by shifting workloads from heavily loaded to lightly loaded machines, ensuring that no two machines are idle at the same moment and that various workloads are ready to be processed. [39]. In [40] authors arrange jobs to relevant machines and machine allocation to operators in order to reduce unbalanced workloads among operators. However, there are several continuing activities in the evolutionary algorithm for load balancing implementation [41]. In [42] authors updated the genetic set of rules for balancing load in a distributed system.

15.4.3 LOAD BALANCER

Every server in the Cloud Data Center informed the load balancer about its available resources unit. And then there's the remaining resource unit, which contains the CPU for processing and RAM for storing data. Every CPU and RAM for each individual host is assigned to a set (HS) by the load balancer. Let $HS = (C_1, M_1), (C_2, M_2), (C_3, M_3), \dots (C_n, M_n)$ The collection of HS contains some irregular hosts that do not have enough resources to complete the specified task in the allotted period. A host with a very fast processing CPU unit and less memory unit is deemed, whereas a host with more memory unit and less CPU unit is regarded a less desirable host for processing.

15.4.4 EXERCISES FOR BALANCING THE LOAD

Allocating the tasks to VMs and scheduling it is based on the demands of arranging the cloud computing workload. To balance the load so many activities are involved and the following activities:

Recognition of the Task: In this the resource which is required for the user tasks is scheduled for execution on a Virtual Machine.

Recognition of the Resources: The status of resources on a virtual machine will be checked in this. It provides information about unallocated resources as well as information about VM resource use. And computed information may be used to identify the state of VM. It also indicates whether it is balanced, overloaded, or underloaded in relation to a particular threshold value.

Scheduling: After completing the above activities the scheduling of task on VMs to an appropriate resource is done by scheduling algorithm.

15.4.5 ALLOCATION OF RESOURCES

This activity now executes the resources assigned to the scheduled tasks. There are several allocation or scheduling options. Policies are being proposed. Essentially, scheduling is necessary to speed up process execution, and allocation strategy is essential for effective resource management and enhancing resource performance.

15.4.6 MIGRATION

It is a critical stage in the cloud load balancing process. There are two kinds of migration: virtual machine migration and task migration. The relocation of a virtual machine from one physical host to another is referred to as VM migration. There have been several ways offered for migration. The most efficient migration is one that results in an efficient load balancing approach. And it has been discovered that task migration is more expensive and time consuming than the VM migration method, hence task migration is currently in use (Figure 15.2).

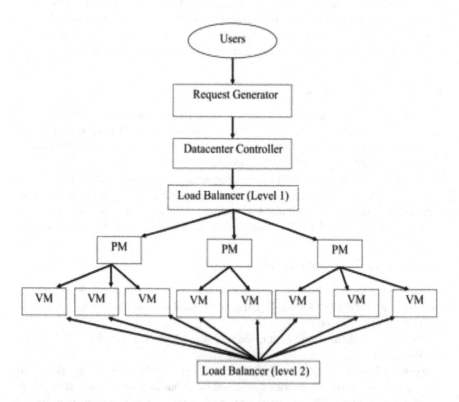

FIGURE 15.2   Load balancing architecture.

15.5   RELATED WORK

In cloud computing, a lot of load balancing approaches have been proposed. For last some decades so many works have been done in the cloud computing environment. There are so many things which have to be scheduled such as tasks, virtual machines, computational units, allocation and management of resources, power consumption, load unbalance. Due to the nature of cloud computing between cloud service providers and cloud service providers, after analysing the existing overview literature, researchers lack proper classification of various methods. This section provides a complete overview of existing works in the field of load balancing in cloud computing. Gomi et al. presents an overview of load balancing algorithms in cloud computing [5]. The authors classified task scheduling and load balancing algorithms into seven different categories, including a two-tier load balancing architecture. The Hadoop map reduction in load balancing, proxy-based load balancing, load balancing with the help of natural phenomena, targeted load balancing for applications, general load balancing, network awareness load balancing and load balancing of specific workflows are divided into two areas in the literature according to the state of the system and the initialization of the process given by Afzal and Kavitha [30]. Various algorithms are summarized in each category and listed. At the same time, new revised load balancing methods are identified during the research the author, Milani and Navimipour [7] divided the load balancing algorithms into three parts, that is, static, dynamic, and hybrid. The authors raised questions about load balancing and addressed major issues related to importance, expected performance levels, roles, and performance balancing and load balancing issues. This was done to use Boolean operations on the search string to find the most relevant search queries from different publication sources, and the selection standard steps are completed through Quality Assurance and Control (QAC). The survey checks only fixed amount of quality of service (QoS) indicators in its operations, that is, durability, scalability of the tasks, utilization of resources, duration, time required for migration, performance, energy consumption and was considering other important QoS indicators (such as cost of migration, interruption in the services, Degree unbalance, percentage of rejection. This survey fills the gap in the selection of analysis indicators.

Kalra and Singh [8] made a comparison. They considered five most important methods, that is, ant colony optimization (ACO), genetic algorithm (GA), particle swarm optimization (PSO), algorithm alliance champion (LCA), and Bat algorithm (BA), exploring different programming algorithms for network and cloud computing. In addition, this review only focuses on evolutionary algorithms, and does not make the classification broad. The load balancing algorithms are divided into three categories by Mesbahi and Rahmani [9] based on artificial intelligence (AI) architecture, they studied different designs and basic requirements to achieve the basic requirement for load balancing. As the authors raised the key issue of developing a load balancing algorithm. In addition, the authors made a conclusion based on the researched algorithms that exhibit the best dynamics, distribution, and noncooperation behaviour [10]. Same as above proposed algorithms the existing load balancing algorithms are divided into dynamic and static algorithms. It also discovered the problem of finding a solution to the load balancing problem. Such as performance, response time of server, migration time, speed of execution etc. The conclusion of the paper is that there is a trade-off between the indicators. The limitation of this document is it compares only eight load balancing algorithms among a large number of algorithms [11]. Provides a complete description of the load balancing algorithm. Various load balancing methods are divided into static and dynamic according to the system state, and divided into homogeneous and heterogeneous according to the homogeneity of virtual machine types. Performance indicators are also used to classify load balancing methods. Algorithms are also discussed. This paper discusses the load balancing problem of the same machine in the work plan of a workshop, where tasks are divided into batches [12].

15.6 METHODOLOGY

We plan to use Grey Wolf Optimization, a novel heuristic swarm intelligent optimization algorithm. The GWO is modelled after the natural leadership structure and hunting mechanism of grey wolves. As the apex predator in the food chain, the wolf has a remarkable capacity to catch prey. Wolves like social interactions, and there is a rigorous social structure within the pack. For replicating the leadership structure, four sorts of grey wolves are used: alpha, beta, delta, and omega. To conduct optimization, three primary processes of hunting are implemented: looking for prey, surrounding prey, and attacking prey. The wolves are divided into four types of wolves: alpha, beta, delta, and omega, where alpha is the best individual, beta is the second best individual, and delta is the third best individual, and the rest of the wolves are considered as omega. The optimization in the GWO is directed by alpha, beta, and delta. They then led the other wolves to the finest place in the hunting field.

It may be visible from Figure 15.4 that the wolf on the role (X, Y) can relocate itself role across the prey in step with above updating formulation. Although the figure show simplest suggests that the seven positions of the wolf are circulate to, through adjusting the random parameters, that is, C and A we can take the wolf to relocate itself to any behaviour with inside the area close to the prey. The GWO assumes that the behaviour of alpha, beta, and delta is most likely optimal. The satisfactory person, second satisfactory person, and third satisfactory person gained up to now are documented as alpha, beta, and delta in the new release looking procedure. Other wolves labelled as omega, on the other hand, shift their positions in accordance with the positions of alpha, beta, and gamma. To change the location of the wolf omega, apply the following mathematical formula.

15.6.1 OVERVIEW OF ALGORITHM

From engineering to economics, and from vacation planning to the internet, effective usage can be seen everywhere. Because money, resources, and time are constantly limited, making the most use of what is available is critical. GWO is a metaheuristic (framework to develop problem independent optimization algorithms) Grey wolf hunting methods influenced the algorithm. The GWO algorithm is modelled after the natural leadership structure and hunting mechanism of grey wolves. For replicating the leadership structure, four sorts of grey wolves are used as we know that. Furthermore, the three basic processes of hunting are implemented: seeking for prey, surrounding prey, and attacking prey (Figures 15.3 and 15.4).

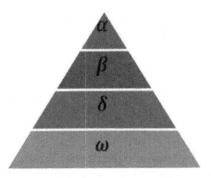

FIGURE 15.3   Grey wolf hierarchy model.

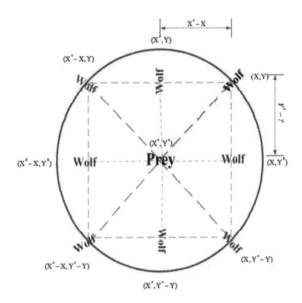

FIGURE 15.4   Position vector and next fit value.

Hunting consists of three basic steps: seeking for prey, surrounding prey, and attacking prey.

Searching for Prey: Tracking, pursuing, and approaching the prey are all part of this process.

Encircling Prey: Pursue the prey, encircle it, and annoy it until it stops moving.

Attacking Prey: Strike towards the prey.

15.6.2   FITNESS FUNCTION

The fitness function is an objective function that summarises, as a single figure of merit, how close a particular design solution is to attaining the defined objectives. Algorithms employ fitness functions to drive simulations toward optimal design solutions.

$$\text{Fitness} = (\alpha * (1/F)) + (1 - \alpha) * (\text{Avgload} - \text{load})$$

where F is the execution time of server s, load is the current load on server s, Avg load is the average load of the system, and α is a random value in range (0, 1).

15.6.3 Algorithm

The behaviour of the wolf pack in the pursuit of prayer, where there is the simplest management class in alpha (α), beta (β), delta (δ), omega (ω).

15.6.3.1 Encircling Prey

The algorithm mimics the part around the victim where the wolves place themselves according to the victim, we use the following statistics:

$$D = |CX(t)pX(t)| \tag{15.1}$$

$$X(t+1) = X(t)pAD \tag{15.2}$$

When D is the calculated distance between the victim and the wolf, the current iteration, $X(t)$ p is the current location of the prey, and $X(t)$ is the position of the wolf. A and C are coefficients and are calculated as follows:

$$A = 2ar_1a \tag{15.3}$$

$$C = 2r_2 \tag{15.4}$$

15.6.3.2 Hunting

Other omega wolves (the last place in the wolf population) stop when they surround the prey according to their alpha, beta, and delta distances. We use Equations (15.1) and (15.2) to calculate the positions between the wolf and the omega and to find the omega position from the ratio of three positions.

$$D_\alpha = |C1 * X(t)_\alpha X(t)|$$
$$D_\beta = |C2 * X(t)_\beta X(t)|$$
$$D_\delta = |C3 * X(t)_\delta X(t)|$$
$$X1 = X(t)_\alpha A1 * D_\alpha$$
$$X2 = X(t)_\beta A2 * D_\beta$$
$$X3 = X(t)_\delta A3 * D_\delta$$
$$X(t+1) = (X1 + X2 + X3)/3$$

Attacking Prey(exploitation): When the prey stops moving, the grey wolf concludes the hunt by attacking it, and a mathematical model is used to minimise the value of the vector A which is a random value in interval [2a, 2a], where a is reduced from 2 to 0 over time. $|A| < 1$ forces the wolves to strike the prey (exploitation).

Searching for Prey(exploration): If $|A| > 1$ then grey wolves are going to be separate from the animal in order to hopefully find a suitable prey (exploitation). One of the GWO item which likes to check exploration is vector C. It contains a random value between [0, 2]. $C > 1$ emphasizes the attack while $C < 1$ emphasizes the attack.

Initialize the position of n grey wolves, where n is the number of search agents.
Calculate coefficients
Calculate the fitness of grey wolves using a fitness function.
Find Alpha = best fit grey wolfs,
Beta = second best and
Delta = third best.
While (the end criterion is not satisfied)
For every wolf
Update the position of wolf using Equations (15.1–15.3)
End For
Calculate All grey wolfs fitness.
Update Coefficients
Update Alpha, Beta and delta.
End While
Return Alpha.

15.6.3.3 Implementation of Proposed Algorithm

Reduction in the makespan can be done by allocating tasks to an appropriate container. To imitate the load balancing using GWO, the following implementation and modification can be done in traditional GWO. Server space mimics the actual possible server location defined within different axis. To follow the leadership hierarchy, a different search server is selected from the server space, which is considered to be the search agent to best fit for the upcoming job. All the selected population or server try to find the best fit server according to fitness function, which is dependent on server load and makespan. The selected server is divided among different levels according to the grey wolf leadership hierarchy. This solution can be further improvised by moving the omega wolf to a different position to attack the job. The omega wolf position is changed according to Equation (15.3) but to normalize these positions to server space we try to find the minimum possible distance of the server from the current omega wolf position so as to select a new set of servers. Accordingly on changing omega position, all the servers in consideration within our population can again check for best fit and on the other hand heads toward hunting the job (Figure 15.5).

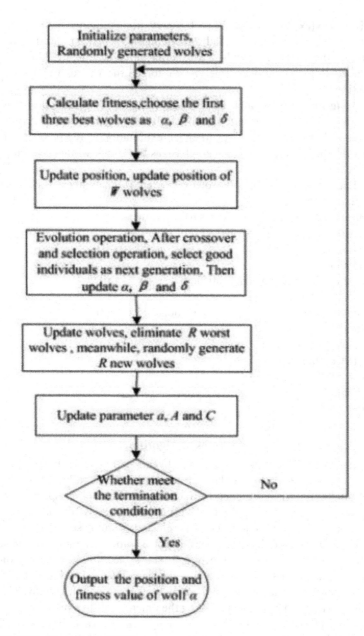

FIGURE 15.5   Flow chart of GWO.

15.7   RESULT ANALYSIS

We regard the system to be heterogeneous in this case, meaning that there are several servers. We suppose that the containers processing capacity ranges from 2,500 to 4,000 MIPS and that the RAM size ranges from 2 to 8 GB. Using a Poisson

Processing Capacity	2400-4000 MIPS
Memory Capacity	2GB, 4GB, 8GB
Arrival Time	[0-1000] ,using Poisson Distribution
Task Size	5000-1000 MI
Number of Tasks	1000,2000,3000
Number of Container	10,20,30,40,50

FIGURE 15.6   Parameter assumed for given system model.

distribution, we suppose that a tasks arrival time ranges from 0 to 1,000. The number of jobs considered is 1,000, 2,000, and 3,000, respectively, while the number of containers considered is 1,050. The number of jobs considered is 1,000, 2,000, and 3,000, respectively, while the number of containers considered is 1,050. Tasks range in size from 5,000 to 10,000 MI. Figure 15.6 lists all of the parameters that are considered in our model. The number of iterations for the GWO method is determined by completing 1,000, 2,000, and 3,000 jobs on 10, 20, 30, 40, and 50 containers, respectively. After some time, both makespan and load values were constant. As a result, for our suggested method, we set the iterations number to 1,000.

For the GFO, we use a population size of 100. After taking all of the considerations into consideration, the outcome of our algorithm is compared to that of the PSO and genetic algorithms. And, when compared to the other two, the GWO produced better outcomes. In the event of varying numbers of containers, the makespan computed using GrWO is smaller than that computed using genetic algorithm and PSO. Similarly, when the number of containers is reduced, the load fluctuation is greater utilising GWO. In comparison to the genetic algorithm and PSO, as the number of containers grows, the load variance in GWO decreases and performance improves. The outcomes are depicted in the diagrams below. The graph shows that as the number of containers increases, load variation and makespan decrease (Figures 15.7–15.12).

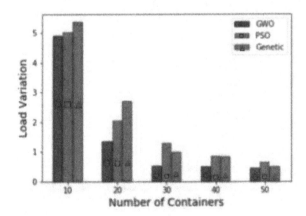

FIGURE 15.7   Load variation with 1,000 tasks.

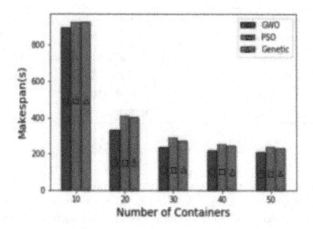

FIGURE 15.8   Makespan with 1,000 tasks.

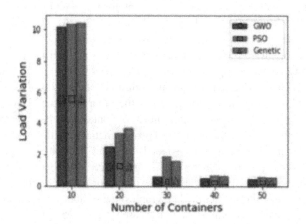

FIGURE 15.9   Load variation with 2,000 tasks.

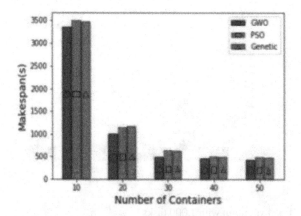

FIGURE 15.10   Makespan with 2,000 tasks.

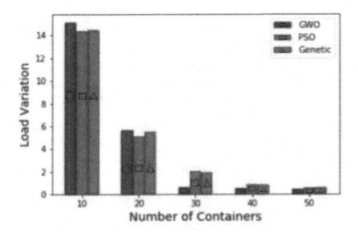

FIGURE 15.11   Load variation with 3,000 tasks.

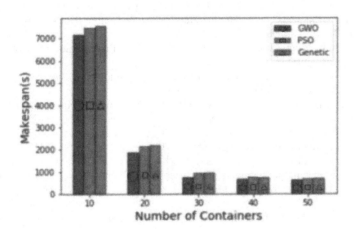

FIGURE 15.12   Makespan with 3,000 tasks.

15.8   DISCUSSION

In this, we mainly focused on the workload distribution which is equally distributed between all containers and minimizing the makespan. There are so many methods have been proposed by various researchers for measuring the loads. In this paper, we developed a GWO algorithm for balancing the load and makespan reduction. Then, GWO-based algorithm performed well as compared with GA- and PSO-based algorithms. The result of GWO is less than GA and PSO. Initially, the load differences in the GWO algo are higher, but when the vessel volume increases, the load decreases and the GWO works better in case of makespan and load balancing than GA and PSO.

REFERENCES

1. Z. M. Gao, J. Zhao, An improved grey wolf optimization algorithm with variable weights, *Computational Intelligence and Neuroscience*, vol. 2019,13 pages, 2019. https://doi.org/10.1155/2019/2981282.
2. S. Petrovic, C. Fayad, A genetic algorithm for job shop scheduling with load balancing. In: Zhang, S., Jarvis, R. (eds) *AI 2005: Advances in Artificial Intelligence. AI 2005*. Lecture Notes in Computer Science, vol 3809. Springer, Berlin, Heidelberg, (2005). https://doi.org/10.1007/11589990_36.
3. R. Swathy, B. Vinayagasundaram, G. Rajesh, A. Nayyar, M. Abouhawwash, M. Abu Elsoud, Game theoretical approach for load balancing using SGMLB model in cloud environment, *PLoS One* 15(4) (2020) e0231708.
4. S. Karimi, Z. Ardalan, B. Naderi, M. Mohammadi, Scheduling flexible job shops with transportation times: Mathematical models and a hybrid imperialist competitive algorithm. *Applied Mathematical Modelling*, Vol 41, (2017) 667-682. ISSN 0307-904X. https://doi.org/10.1016/j.apm.2016.09.022.
5. E. J. Ghomi, A. M. Rahmani, N. N. Qader, Load balancing algorithms in cloud computing: a survey, *J Netw. Comput. Appl.* 88 (2017) 50–71.
6. E. Shamsinezhad, A. Shahbahrami, A. Hedayati, A. K. Zadeh, H. Banirostam, Presentation methods for task migration in cloud computing by combination of Yu router and Post-Copy, *IJCSI International Journal of Computer Science Issues*, 10 (4) (2013) 98-102.
7. A. S. Milani, N. J. Navimipour, Load balancing mechanisms and techniques in the cloud environments: systematic literature review and future trends, *J. Netw. Comput. Appl.* 71 (2016) 86–98.
8. M. Kalra, S. Singh, A review of meta-heuristic scheduling techniques in cloud computing, *Egypt Inform. J.* 16(3) (2015) 275–295.
9. M. Mesbahi, A. M. Rahmani, Load balancing in cloud computing: a state of the art survey, *Int. J. Mod. Educ. Comp. Sci.* 8(3) (2016) 64.
10. V. R. Kanakala, V. K. Reddy, K. Karthik (2015, March). Performance analysis of load balancing techniques in cloud computing environment. In: *2015 IEEE International Conference on Electrical, Computer and Communication Technologies (ICECCT)*, SVS College of Engineering, Coimbatore, India, pp. 16.
11. J. M. Shah, K. Kotecha, S. Pandya, D. B. Choksi, N. Joshi (2017, May). Load balancing in cloud computing: methodological survey on different types of algorithm. In: *2017 International Conference on Trends in Electronics and Informatics (ICEI)*, Tirunelveli, India, pp. 100107.
12. P. Bruker, R. Schlie, Jobshop scheduling with multipurpose machines, *Computing* 45 (1990) 369–375.
13. W. J. Xia, Z. M. Wu, An effective hybrid optimization approach for multi-objective flexible jobshop scheduling problems, *Comput. Ind. Eng.* 48 (2005) 409–425.
14. S.-S. Kim, J.-H. Byeon, H. Yu, H. Liu, Biogeography-based optimization for optimal job scheduling in cloud computing, *Applied Mathematics and Computation*, vol. 247, (2014) 266–280. https://doi.org/10.1016/j.amc.2014.09.008.
15. I. C. Choi, D. S. Choi, A local search algorithm for jobshop scheduling problems with alternative operations and sequence dependent setups, *Comput. Ind. Eng.* 42 (1) (2002) 43–58.
16. J. Gao, M. Gen, L. Sun, Scheduling jobs and maintenances in flexible job shop with a hybrid genetic algorithm, *J. Intell. Manuf.* 17 (4) (2006) 493–507.
17. N. Imanipour, S. H. Zegordi, A heuristic approach based on Tabu search for early/tardy flexible jobshop problems, *Sci. Iran.* 13 (1) (2006) 113.
18. P. Fattahi, M.S. Mehrabad, F. Jolai, Mathematical modeling and heuristic approaches to flexible job shop scheduling problems, *J. Intell. Manuf.* 18 (3) (2007) 331–342.
19. P. Fattahi, F. Jolai, J. Arkat, Flexible job shop scheduling with overlapping in operations, *Appl. Math. Model.* 33 (7) (2009) 3076–3087.

20. C. Ozguven, L. Ozbakir, Y. Yavuz, Mathematical models for jobshop scheduling problems with routing and process plan flexibility, *Appl. Math. Model.* 34 (6) (2010) 1539–1548.

21. C. Ozguven, Y. Yavuz, L. Ozbakir, Mixed integer goal programming models for the flexible jobshop scheduling problems with separable and nonseparable sequence dependent setup times, *Appl. Math. Model.* 36 (2) (2012) 846–858.

22. V. Roshanaei, H. ElMaraghy, A. Azab, Mathematical modelling and a meta-heuristic for flexible job shop scheduling, *Int. J. Prod. Res.* 51 (20) (2013) 6247–6274.

23. P. Brandimarte, Routing and scheduling in a flexible job shop by Tabu search, *Ann. Oper. Res.* 41 (14) (1993) 157–183.

24. H. Chen, J. Ihlow, C. Lehmann, A genetic algorithm for flexible jobshop scheduling, in: *IEEE International Conference on Robotics And Automation*, 2, Detroit, 1999, pp. 1120–1125.

25. G. Zhang, L. Gao, Y. Shi, An effective genetic algorithm for the flexible jobshop scheduling problem, *Expert Syst. Appl.* 38 (2011) 3563–3573.

26. M. Mastrolilli, L.M. Gambardella, Effective neighbourhood functions for the flexible job shop problem, *J. Sched.* 3 (1) (2000) 320.

27. M. Yazdani, M. Amiri, M. Zandieh, Flexible jobshop scheduling with parallel variable neighborhood search algorithm, *Expert Syst. Appl.* 37 (2010) 678–687.

28. W. Xia, Z. Wu, An effective hybrid optimization approach for multi-objective flexible jobshop scheduling problems, *Comput. Ind. Eng.* 48 (2) (2005) 409–425.

29. J. Li, Q. Pan, Chemical reaction optimization for flexible jobshop scheduling problems with maintenance activity, *Appl. Soft. Comput.* 12 (2012) 2896–2912.

30. S. Afzal, G. Kavitha, Load balancing in cloud computing–A hierarchical taxonomical classification, *J. Cloud Comput.: Adv., Syst. Appl.* 8 (2019) 22.

31. B. Naderi, A. Ahmadi Javid, F. Jolai, Permutation flowshops with transportation times: mathematical models and solution methods, *Int. J. Adv. Manuf. Technol.* 46 (2010) 631–647.

32. V. A. Strusevich, A heuristic for the two machine open shop scheduling problem with transportation times, *Discrete Appl. Math.* 93 (1999) 287–304.

33. J. Hurink, S. Knust, Tabu search algorithms for jobshop problems with a single transport robot, *Eur. J. Oper. Res.* 162 (2005) 99–111.

34. M. A. Langston, Interstage transportation planning in the deterministic flowshop environment, *Oper. Res.* 35 (1987) 556–564.

35. B. Naderi, M. Zandieh, A. Khaleghi Ghoshe Balagh, V. Roshanaei, An improved simulated annealing for hybrid flowshops with sequence dependent setup and transportation times to minimize total completion time and total tardiness, *Expert Syst. Appl.* 36 (2009) 9625–9633.

36. M. Boudhar, A. Haned, Preemptive scheduling in the presence of transportation times, *Comput. Oper. Res.* 36 (2009) 2387–2393.

37. J. Wu, Y. Li, C. Peng, & Z. Wang, Wideband and low dispersion slow light in slotted photonic crystal waveguide, *Optics Communications*, 283(14) (2010) 2815-2819.

38. D. Kranzlmüller, P. Heinzlreiter, H. Rosmanith, & J. Volkert, Grid-enabled visualization with gvk, In European Across Grids Conference, Springer, Berlin, Heidelberg, 2003, pp. 139-146.

39. A. Y. Zomaya & Y. H. Teh, Observations on using genetic algorithms for dynamic load-balancing, *IEEE transactions on parallel and distributed systems*,12(9) (2001) 899-911.

40. D. H. Moon, D. K. Kim, & J. Y. Jung, An operator load-balancing problem in a semi-automatic parallel machine shop, *Computers & Industrial Engineering*, 46(2) (2004) 355-362.

41. W. A. Greene, Dynamic load-balancing via a genetic algorithm, in: *Proceedings 13th IEEE International Conference on Tools with Artificial Intelligence*, ICTAI 2001, (2001) pp. 121-128.

42. C. F. Huang, H. W. Lee, & Y. C. Tseng, A two-tier heterogeneous mobile ad hoc network architecture and its load-balance routing problem, *Mobile networks and applications*, 9(4) (2004), 379-391.

16 A COVID-19 Vaccine Notifier Android App "CoWin Mitra"

Vikas Kumar Patel and Anshul Verma
Banaras Hindu University

Pradeepika Verma
Indian Institute of Technology Patna

CONTENTS

16.1 INTRODUCTION

On January 16, 2021, India has started its COVID-19 vaccination drives with two vaccines named Covishield (AZD1222) with efficiency about 80.6% and Covaxin (BBV152) with efficiency about 70.4% [1]. Initially, approximately 12,000 private and central government hospitals have been chosen as vaccination sites. Government of India launches a web portal for free/paid COVID-19 vaccination for the people of India with different age groups (18–44 and 45+). User must first register and take an appointment for vaccine through CoWIN web portal [2]. When there were very few centers with limited vaccination dose, it had become very hard for the user to get an appointment. They always kept their eyes on the web portal for vaccine availability.

This problem can be solved using a notification app that notifies the user about a particular event or according to user need. Since the government of India makes the vaccination service APIs [3] publicly, which means any third party can read the details of vaccination centers, and other vaccine-related details, many android apps and web apps have been developed to solve this problem, but none of them provide an in-app appointment booking feature. They notify the user about the available vaccine and put a link to the CoWIN web portal [2]. To book the appointment, users have to remember the details such as pin code and center name. The proposed app named "CoWin Mitra" uses these APIs to fetch the details of centers in a specific district and regularly checks for available vaccination slots in that district. If the app found new slots, it notifies to the user about the slot with center details.

Since the API service (CoWIN) provided by the government (API Setu [4]) is limited, the only authorized person can make an appointment for an individual. To overcome this limitation, the proposed app opens the official web portal of CoWIN within it for user registration/slot booking. Users can also get help through the provided information about the centers in the screen for easy-to-use.

This app has been developed using android software development kit (SDK) [5] version 29. To store the information about important entities such as state name, center details, and session details, this app uses a client-side database technology SQLite [6]. Finally, a white box test and a black box test [7] have been performed to test the design, functions, logics, and compatibility of the android app. The aim of this work is to design basic UI, then to add significant functionality, thereafter, to redesign it if needed. The proposed design is then tested for verification and finally implemented. With this approach, errors can be easily detected and corrected at the same time. The main objectives of development process are as follows:

- To fetch and store the APIs data in SQLite database.
- To show filtered vaccination slots based on age-group, dose type, vaccine name.
- To subscribe/unsubscribe to the individual center.
- To provide the facility to users to enable/disable notifications update any time.
- To provide manually refresh the vaccination slot list option to users.
- Notification should be delivered only for subscribed places and with applied filtered.
- Redirect users to the web portal when they click on notifications.

16.2   LITERATURE SURVEY

Several web app and android app have been developed in the past with the idea of notifying the user about available vaccines, scheduling an appointment for vaccination. A few of them are as follows.

VaccinateMe: It is a web app developed by HealthifyMe [8]. It checks for the available vaccine about every 5 minutes and notifies registered users about available vaccines through WhatsApp/telegram. Server-side script is used to send notifications, these scripts run every minute.

Getjab: It is a web app, also named as "CoWIN slot notifier" [9], developed by Azhar, Shyam, Anurag, and Akshay. This web app sends emails regarding available vaccines. Also, they planned to provide SMS support that means user will receive SMSs regarding available vaccines. It checks for the available vaccine on server side and send the email to users who are looking for slots.

CoWIN Vaccine Tracker: The CoWIN Vaccine Tracker [10] android app helps users to find the available vaccines (pin code, district, and age-group wise). User can search for vaccine centers available for both age-group 18+ and 45+. It detects real-time drug availability every 30 seconds and sends alerts. The purpose of making this app is to assist Indians in booking COVID-19 vaccines. The app does not require any special permissions such as storage access, contacts, calls, GPS, and so on. It connects directly with the official CoWIN APIs and provides real-time data according to pin code, district, and age-group.

Vaccine Slot Notifier: The objective of Vaccine Slot Notifier [11] app is to deliver notifications regarding available vaccine slots by district or by pin code. The app only notifies once, if the user wants to get notified again, they must open the app and reenable the notification.

Vaccine Slot Tracker: Vaccine Slot Tracker [12] is an android app that notifies the available vaccine for favorite pin code and favorite district. There is no other filter available in the app according to the age-group or location notification, that is, users are notified for all age-group vaccines and from all centers located at favorite pin code and/or favorite district. It checks for the vaccine every 2 minutes. If the user wants to stop the notifications, they need to remove all favorite places from the list.

Findslot.in: Findslot.in [13] is a website that lists the available vaccine for the user. Just like the official CoWIN portal search feature, this website searches and shows the list of available vaccine doses according to the user's city and age-group or according to the pin code and age-group.

CoWin Vaccine Slot Finder: CoWin Vaccine Slot Finder [14] is an open-sourced project available on GitHub named as CoWinVaccineSlotFinder. This project can be run on Windows, Linux, and macOS. The app store beneficiary details and center details, and run an automated script in CUI (Command user interface). For this, the end-user must have basic knowledge of CUI. Also, the user must edit the configuration file with their details before running the application.

Aarogya Setu: The primary objective of Aarogya Setu [15] mobile application is to inform the users about COVID-19 positive people within range of 500 m–10 km. This app is developed by the Government of India. It requires some special permissions like location access, Bluetooth access, SMS. User can perform self-test using the app to know whether they have symptoms of COVID-19 or not. User can schedule and manage vaccination appointment in the app but they do not get any notifications about available vaccines.

CoWIN Vaccinator App: This mobile application (developed by National Health Portal-MoHFW) [16] is currently meant for CoWIN Facility level users to perform following tasks as Vaccinator, Supervisors and Surveyors: Beneficiary Registration, Beneficiary Verification, Aadhaar Authentication, Vaccination Status.

After reviewing and studying related developments, projects, and apps it seems that all these android apps, web apps, and Windows/ Linux/ MacOS app, have different approach with different functionalities of the project presented, but none of them have introduced all the functionalities simultaneously. Since, all these apps use same backend to get data, most of them just show the information about available vaccination centers. Users have to open the portal through web Brower and then search for the centers' name. Since this process takes some time, the booked slot may be allotted to someone else. The project mentioned below has been developed to fix all these issues occurred during notification generation.

16.3   PROPOSED APPROACH

Software Development Life Cycle (SDLC) [17] is a process to design, develop and test high-quality software. Aim of the SDLC is to build high-quality and more stable software that meets customer requirements within times and cost estimates.

The SDLC defined its phases as demand collection, design, coding, testing and maintenance. It is important to attach to the stages to provide the product consistently.

16.3.1   SDLC MODELS

Various software development life cycle models are specified and designed which are followed over the development process. These models are also known as "software development process models". Every model has its own defined unique processes to build the software.

This project has been developed using iterative model [17]. It is a specialized implementation of a SDLC that aims to an early, simplified implementation at the initial stage, further it becomes progressively more complex and has a wider feature set until the final system is completed. This approach is based on partitioning the software development process of a vast application into tiny pieces. A Project Control List (PCL) [17] has also been developed based on the current known requirements. It contains a chain of all the functions that are available in the provided system.

To develop the android app, each function is selected from the PCL to complete the planning, requirement analysis, UI designing, testing and implementation processes as illustrated in the Figure 16.1. The specified function is further removed from the PCL once these processes for that function have been completed. These processes are repeated until the desired requirements of the software or part of software are achieved.

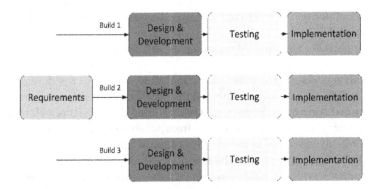

FIGURE 16.1   Iterative model of the proposed app.

16.3.2 PROJECT REQUIREMENTS

In this work, a requirement list has been generated for the smooth and successful development of project and to maintain uniformity in implementations during the development process. The generated requirement list is as follows.

- User Interface to ask for user-permission (Ignore Battery Optimization to run the app in the background).
- Read user state and district
- Add filters – vaccine, age-group, dose
- Recycler – view to show and choose a different available date
- Recycler – view to display center information
- Automatic refresh time interval interface
- Vaccine filters interface
- App setting with the following options:
  - Change automatic refresh time interval
  - Manage vaccine filters
  - Change permission for battery optimization
- An app bar icon to enable or disable notifications
- A refresh icon to manually refresh the list of available vaccines
- A button to change home district, subscribe to all centers of the home district or unsubscribe from all centers of the home district
- Users can get notifications from only the home district
- Users can individually subscribe/unsubscribe to specific centers
- Users will only get notifications from subscribed centers
- A web-view to load the official CoWIN portal in the app when the user selects any center.
- The web-view should contain important information such as center pin code, name, and date. This will help users to easily search for the required center in the portal.
- The user should be redirected to web-view when they click on a specific notification.

Here, this featured record is also introduced as a project control list throughout development process.

Some targets are also covered regarding its efficiency. These are as follows:

- **Accuracy**: As we are using the official data, this project is developed to provide high accuracy strength. All the operations are executed correctly.
- **Planned Approach**: Every step while building this project is well planned and organized. The data are fetched from government official API endpoints (API Setu) and stored in the user device in the form of SQLite database.
- **Reliability**: In this project, the source of the data is from API Setu, which provides services to many government departments, and hence, reliability of this project can be more. Also, due to proposed app design and implementation process, the data are stored in the user device, and hence, it provides more reliable.
- **Fast Operations**: Since the data are fetched from API and stored in a local database, the operations like filtering, sorting, displaying the data become fast.
- **User-Friendly**: The UI of the app is simplified and designed for easy to use. The data representation is almost same as the official CoWIN portal. In this way, the UI is more user-friendly.

16.3.3   E-R Diagram

Entity Relationship Diagram (ER diagram) [18], also called as ERD, is a diagram that describes the relations between entity sets kept in the database. In more simple words, these diagrams can explain the logical structures of the database built on the basis of three basic concepts, properties, entities, and relationships. It consists of different features such as primary key that is represented with a key icon (null with "N") and foreign keys with a pink shape. Every relation is marked with a structural constraint where the number denotes a cardinality ratio, that is, 1:1 (one to one), 1:m (one to many), or m:n (many to many).

In this project 2, ER diagrams are used during the development process. The ER diagram illustrated in Figure 16.2 shows the relationship of the center model with the various models related to the centers. The center model is attached to the district (M: 1), the district model is attached to the state (M: 1), and the session model is attached to the model centers (M: 1).

The ER diagram shown in Figure 16.3 shows the data about user filters for notifications and time intervals to check for available vaccines from the server.

16.3.4   Message Sequence Chart

It represents an attractive visual modality that is widely used in the domain such as telecommunications software to capture system requirements during the initial design phases. A variant of Message Sequence Charts (MSCs) [19] known as sequence diagrams is one of the behavioral diagram types adopted in the unified modeling language [19].

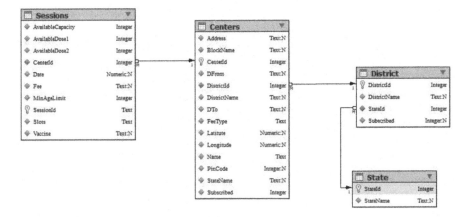

FIGURE 16.2   ER diagram for state model, district model, centers model, and sessions model.

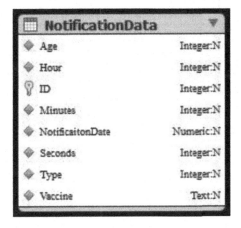

FIGURE 16.3   ER diagram for notification data model.

The purpose of recommending the MSC was to supply a trace language for the description and specification of the communication behavior of system components and their environment through message interchange. Since communication behavior in MSC is presented in a very intuitive and transparent manner, especially in graphical representation, it is quite easy to implement and understand. A basic MSC shows interaction between system components (also known as instances). Every vertical line in MSC describes an instance, and the horizontal arrows show messages from one instance to another.

Two MSCs describe main messages exchanged between the modules of the project. The MSC in Figure 16.4 illustrates the execution of the app when it is launched by the user. When the app is launched for the first time, the intro screen is appeared. After that it asks to users for permission to ignore battery optimization to run the app in the background. Then, "My Choices" screen appears for filter selection and to set

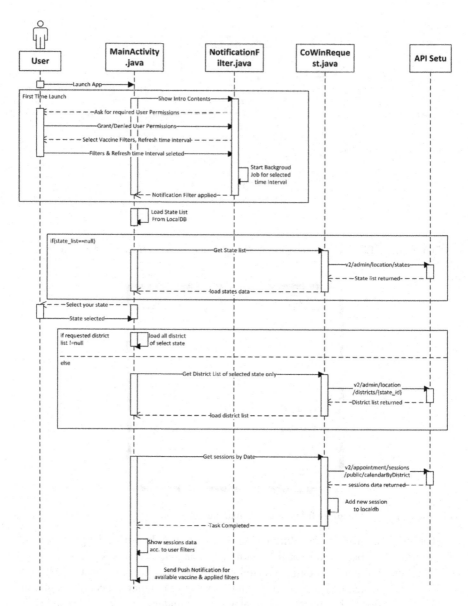

FIGURE 16.4 Message sequence chart for the first run and normal launch of app.

automatic refresh time interval. A background job service start running for every refresh time interval chosen by the user. Thereafter, a state list is to be loaded in a drop-down text view from the local database. If this list does not exist, the app gets the list of states from the server (API Setu [4]). In this sequence, this app asks the user to select the belonging state, after selecting state app checks whether the list of districts for the selected state is stored in the local database or not, if it is, then the data are to be parsed from the local database else the list of districts is to be downloaded

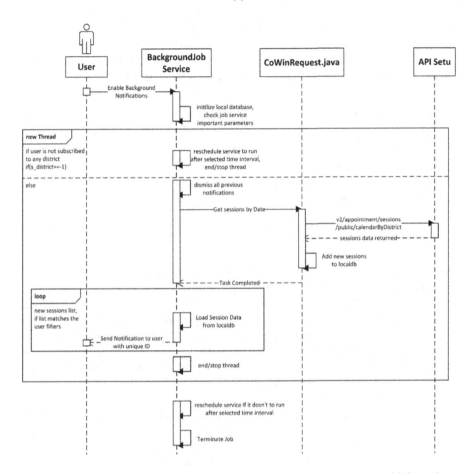

FIGURE 16.5 Message sequence chart describes the data flow of background job service.

from the server. Next, the list of sessions (vaccine availability) is downloaded from the server for the upcoming 7 days. After successfully downloading the sessions list, it is stored in the local database to be presented to the user. If the session list has any vaccine availability according to the user filter, a notification is also be appeared.

The MSC in Figure 16.5 illustrates the working of background job service. This service is controlled by the user. When the user enables notification, they start a service that runs every x second (x is the automatic time interval selected by the user). This service first initializes the database and checks whether the parameters of the service which include the user filters for the vaccine. Since some manufactures use their algorithms to stop the background process and save the battery life, a new separate thread is initialized and any long-running process is included in that thread. In this thread, the app first checks whether the user has subscribed to any district or not; if the user does not subscribe to any district yet, the thread is terminated and service is rescheduled. If a user has subscribed to any district, then first of all any other notifications shown to the user are dismissed. (Since the old notifications are not useful anymore.) After that, the app downloads a list of sessions (vaccine availability) that is downloaded from the server for the upcoming 7 days. The downloaded session list stores in a local database. If any session is found according to the user filter, the notification for each session with a unique notification ID is triggered. In the last, the thread and job service stop after rescheduling it for the next time interval.

16.4 IMPLEMENTATION

16.4.1 TECHNOLOGIES USED

Various server-side and client-side machinery/automation are available to develop this type of project. The technologies used for developing the application are mentioned briefly as follows.

16.4.1.1 Server-Side Technologies

16.4.1.1.1 API Setu

API setu [4] is an API platform to enable swift, transparent, safe, and reliable information sharing across applications and to promote innovation.

An "Open API" policy was notified by the Ministry of Electronics and IT (MeitY) in July 2015 with an objective to develop an open and interoperable platform to enable seamless service delivery across government. It assumes that to provide impetus to the policy implementation, an API platform to act as an API Marketplace cum Directory Portal is necessary. Hence, MeitY has initiated, API Setu also known as the Open API Platform project in March 2020. This platform aims to bring the policy into realization.

More than 150 major central and state government departments are already available on the API Setu platform and provide access to about 600 APIs for various data points such as driving license, vehicle registration, PAN, CBSE, e-District in MeitY's DigiLocker. In this project "CoWIN Public APIs" Version 1.3.1, OAS3 is used to get the vaccine data.

The following API endpoint is used in this project:

1. Metadata APIs
   a. v2/admin/location/states
   b. It is used to get the list of states in India.
   c. v2/admin/location/districts/{state_id}
      This API endpoint is used to get the list of districts of particular state. The state_id is unique ID of state which can be obtained through state metadata API.
2. Appointment Availability API
   a. v2/appointment/sessions/public/calendarByDistrict
      It returns the vaccination sessions by district for 7 days from the selected date.

There are several other APIs available such as to get vaccination sessions by pin code for 7 days or to get vaccination centers by location (latitude and longitude), but in this project, only the mentioned APIs endpoint are used. All these APIs returns the data in JSON format.

16.4.1.2   Client-Side Technologies

16.4.1.2.1   Java

Java [20] is an object-oriented, high-level, class-based programming language that has been built to have as few implementation dependencies as possible. It is a general-purpose programming language intended to let application developers write once, run anywhere, means compiled Java code can run on all platforms that support Java without the need for recompilation.

In this app, Java language with version 11.0.8 is used to implement the different modules of proposed app such as local database parts, fetching data from the API Setu, and all other farcicalities that this app can be implemented.

16.4.1.2.2   XML

Extensible markup language (XML) [21] has some set of rules to encode documents, which are human readable as well as machine readable. It has textual data format.

In this app, XML has been used to code layouts (linear layout, coordinator layout, constraint layout) and UI elements. The version of XML used in this development is XML 1.0.

16.4.1.2.3   SQLite

SQLite [6] is a relational database management system or RDBMS [22] available in a C library. Unlike many other database management systems, it is not a client-server database engine. Rather, it is embedded in the end program. The version of SQLite used in this development is 3.8 and up. (SQLite version has been automatically initialized depending on the user android API level.)

16.4.1.2.4   AndroidSDK

Android SDK [5] has a set of development tools used to develop applications for the android platform. The android SDK includes required libraries, an emulator, a debugger, relevant documentation for the android APIs, sample source code, and tutorials for the android OS. Google releases new versions of android with corresponding SDK and related documents [23] with examples on regular basis. Developers must download and install each version's SDK for the particular devise to be able to write programs with the latest features. The version of android SDK used in this development is android 11 (API level 30).

16.4.1.2.5   Retrofit

Retrofit [24] is a type-safe REST client for android and Java that aims to make it easier to consume RESTful web services [3]. Retrofit version 2 (also called retrofit-2) leverages OkHttp as the networking layer by default and is built on top of it. Retrofit automatically serializes the JSON response using a Plain Old Java Object (POJO) that must be predefined for the JSON structure to be created. It is very useful for handling large data responses sent by the server. The version of Retrofit used in this development is Retrofit 2.9.0.

16.4.2 HARDWARE AND SOFTWARE REQUIREMENTS

This project covers a wide range of devices since the hardware configuration of every device is not the same. The app size and response time may vary for different devices. The device should have at least 20 MB free device storage for installation, with free RAM of 50 MB (this may vary from device to device). The minimum software requirement is android 5.1 (SDK 22), and the recommended software requirement is android 10 (SDK 29).

16.4.3 METHODOLOGY

This project is developed via multiple steps. The significant steps of the proposed methodology are listed below:

1. Installing android studio with android virtual device (AVD) from the official android developer page
2. Installing the required/latest android SDK
3. Creating a new project and initializing Git for the project directory
4. Selecting project template from template gallery (android black activity template is used in this project)
5. Provide details such as App name, package name.
   a. App name: **CoWin Mitra**.
   b. Package name: **in.cyberguardian.cowinmitra**.
6. Adding/updating new dependencies required for the project for example retrofit, firebase
7. Commit the changes to Git
8. Run and deploy the project

The flow chart of development of the proposed app is illustrated in Figure 16.6.

16.5 RESULT

The snapshots of the developed app are shown describing the functionalities of the project.

The homepage screen consists of some introductory information about the app. This screen can be visible only once (at the first launch of the app) to the user describing the objectives of the app. On clicking on the "Get Started" button, user has been asked to select vaccination filters and automatic refresh time intervals to check for updates to deliver notifications. By default, the filters and automatic refresh time interval values are shown in the "My Choices" screen. Users can scroll through the filters to change them. This screen can also be accessed through the app setting screen and choosing option either "Change Automatic Refresh Time" or "Manage Vaccine Filters." Clicking on the "Next" button located at the downside right of

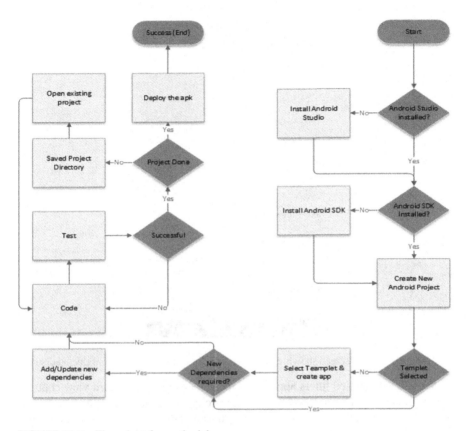

FIGURE 16.6 Flow chart for methodology.

the screen, the selected settings can be saved and users are redirected to the "Main Screen."

The Main Screen (Figure 16.7) is dynamic and responses according to user activity. On the top of the screen, there is an app bar that contains the app name, a refresh button (to refetch the data from the server and to show notifications if any vaccine is available), a notification button (to start/stop automatic checking for the vaccine), and a three-dot menu (app setting and app sharing options). When the app is launched the first time or if the home district is not selected, the user experiences the blank data in the state and district drop-down text view. Below the district drop-down, three chips are placed for filtering the results. When user selects the state and district for the first time, the app marks the selected district as the home district. A log text about the subscribed centers appears in between the district drop-down and filters chips with a notification filter button. This notification button can be used to change the home district, to subscribe to all centers of a district, and to unsubscribe from all centers of a district. Below the filter chips, the date-wise data of the vaccine with numbers of the available vaccine on that day display in recycler-view with the help of card-view. When the user clicks on any given date, the state and district drop-down text view and log text with notification button disappear and the list of centers in the

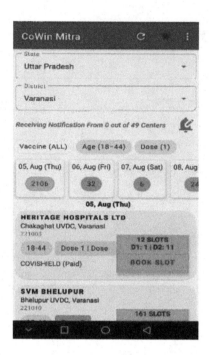

FIGURE 16.7   Main screen showing available doses in different centers for selected district.

sorted order of applied filters appears on the screen. The center card view contains information like center name, pin code, age-group, available dose type (dose 1 or 2), vaccine name, a button on top to subscribe/unsubscribe from the center, a clickable card view that has the numerical data about vaccine doses which has a total number of vaccine available, dose 1 (D1) data, dose 2 (D2) data, and a button "Book Slot" to open registration portal in the app (Figure 16.7).

The In-App vaccine registration screen (Figure 16.8) has an app bar with title in the format: <app name> | <Centre Name>, subtitle contains the center pin code. At the right-top in the app bar, a button is placed to open the CoWIN portal through the user browser.

The notifications (Figure 16.9) can be delivered for every time interval chosen. Title of the notification contains the center name, and the body contains the other details such as when the notification is created, name of vaccine, date, dose type, minimum age-group, district, and pin code.

16.6   TESTING

In the field of android development, many methods are available to test the app. During this app development process, the following testing has been performed:

16.6.1   BLACK BOX TESTING

Black box testing [7] is a method of software testing that examines the functionality of an application without peering into its internal structures or workings. This

FIGURE 16.8   In-app vaccine registration.

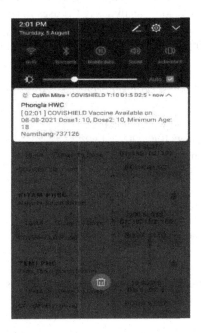

FIGURE 16.9   Showing push notifications about available vaccine.

method of testing can be applied virtually to every level of software testing that are unit, integration, system, and acceptance. In black box testing, the designs, implementation, and internal structure of the app are hidden for the tester.

For this project, 12 android users having android devices by different manufactures have been selected. In the beginning, the users are asked to install/update/uninstall the app, then they are instructed to check the functionalities and user-interface of the app. Three users have raised an issue with the user interface due to small screen size (icon color does not change in dark mode), this bug has been fixed and the further processes are taken place.

In the final phase of black box testing, all testers are requested to revisit the entire application and each functionality. The app uses traces, performance index, app response time, app crashes, and many other data are recorded with the help of Google analytics and firebase crash report along with saving user's feedback. In this way, application's design and implementation parts have been improved.

16.6.1.1   Robo Testing

Robo testing (robotic testing) [25] is a black box testing method to test the user interface, program logics, and other functions of mobile app by AI. In this project, robo testing of Firebase test lab is used. This test is available with two variations. First one is recording robo script to guide robo test according to script. In this, a predefine script is generated by app developer to test specific part of app and the robo follow generated script and test the app with different environment. Another variation is unguided robo testing where only executable application is uploaded in the test lab and rest of things like testing each and every button, entering valid/invalid inputs, closing, and reopening the app many times are done by AI. While testing the app in both variations of method, some analytics (CPU uses in percent, Graphics in frame per seconds, Memory uses in KiB, Network uses in Bytes per second) are recorded which helps in obtaining the results (Pass OR Failed). Test reports and screen captures are given below.

16.6.1.1.1   Test 1: Robo Test on Pixel 3 Device (Android API Level 28)

- Test method: robo test with guided script
- Date and time of test: 10/07/2021, 13:24
- Test duration: 2 minutes 13 seconds
- App start time: 310 ms
- Graphic stats
  - Missed VSync: 2%
  - High Input latency: 4%
  - Slow UI thread: 3%
  - Show draw commands: 1%
  - Slow bitmap uploads: 0%
- CPU, graphics, memory, and network utilizations are shown in Figure 16.10.

16.6.1.1.2   Test 2: Robo Test on Pixel 3 Device (Android API Level 28)

- Test method: robo test without guided script
- Date and time of test: 14/10/2021, 20:41
- Test duration: 1 minute 21 seconds
- App start time: 230 ms
- Graphic stats
  - Missed VSync: 0%
  - High input latency: 1%
  - Slow UI thread: 0%
  - Show draw commands: 0%
  - Slow bitmap uploads: 0%
- CPU, graphics, memory, and network utilizations are shown in Figure 16.10.

TABLE 16.1
Accessibility Report of Test 2

Issue Types	Warnings	Minor Issues	Tips
Touch target size	0	0	0
Low contrast	0	0	0
Content labelling	0	0	0
Implementation	0	0	0

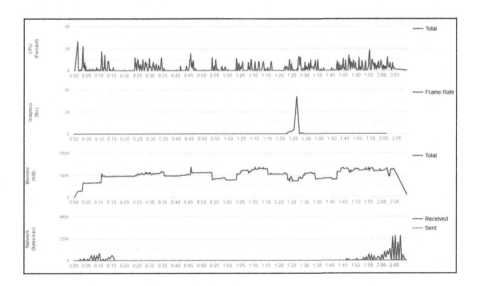

FIGURE 16.10   Performance graph of app in Robo Test 1.

With the above test report, we can conclude that overall test can be considered as pass (both in test 1 and test 2). Some minor changes in app have been made after reviewing the report of test 1 UI performance, which results in high UI performance as mentioned in test 2 (Figures 16.11–16.13).

16.6.2   White Box Testing

White box testing [7] is a method of software testing that tests the internal structures or functionality of an application. In this testing, the tester gets to know about the implementation, design, and internal structure of the project.

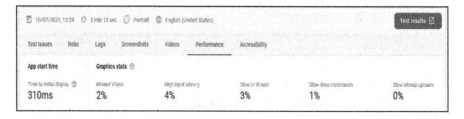

FIGURE 16.11   App launch time and UI performance of app in Robo Test 1.

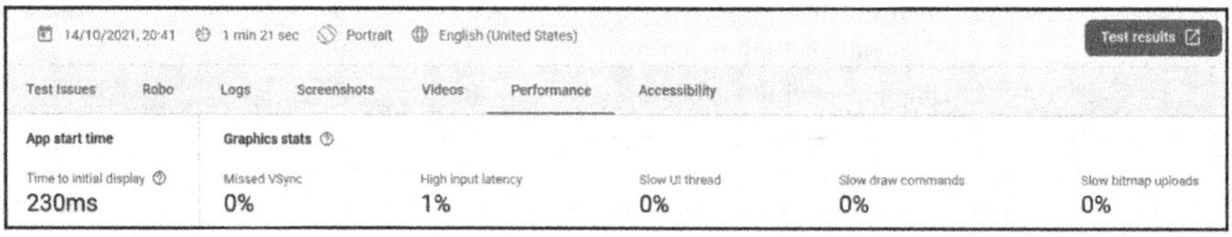

FIGURE 16.12   App launch time and UI performance of app in Robo Test 2.

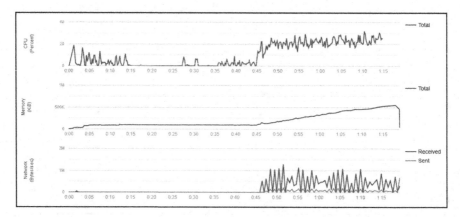

FIGURE 16.13   Performance graph of app in Robo Test 2.

This type of testing is done by developers only. Unit testing (a kind of white box testing) is done in this project. In unit testing, each and every unit of code is tested separately and finally integrated. The test includes the following types of source code:

1. Android security loopholes [26]
2. Possible exceptions in codes
3. Deprecated methods
4. SSL certificate loopholes
5. The functionality of loops/conditional statements
6. Activity exportation
7. Proguard rules
8. Properly minifying the source code
9. Testing of each object, class, and method

Several separate test cases are created and tested for each unit of source code. For each test case, the desired output is expected. If the output is not satisfactory, then it causes a bug. Every error has been fixed in the source code, and all units are finally integrated.

16.7   CONCLUSION

On the basis of the result obtained, the developed android application can be considered extinguishable. The app developed can help many people to book an appointment for vaccination within the app. With the help of filters, the user can get the notifications for only selected dose type, age-group, vaccine, and centers. Users can also disable/enable the notification any time from the app, and rest of the settings/filters remain saved in the app. Searching become very easy on the CoWIN portal because information such as pin code and center name are shown in the app bar.

This android application is developed by implementing the proper software engineering methods, iterative model. A requirement list with the proper prototype is also created. Required ER diagram has been developed for all the tables and relationships between each table. The data flow process has been shown with the help of the MSC, and two MSCs are presented in this paper. The flow chart of methodology has built to perform different tasks in a planned way. Also, each part of the requirement list is designed, coded, and tested using suitable testing methods. Corrections, modifications, and improvements are made according to test user feedback. Also, two robo tests with two different methods have been performed, which shows that the app is very light weight and consumes a smaller number of resources.

In the future work, some other features such as showing the result based on pin code, sorting the center list according to user location, showing the data in the map view, would be useful to implement in this project.

ACKNOWLEDGMENT

This work is part of the research work funded by "Seed Grant to Faculty Members under IoE Scheme (under Dev. Scheme No. 6031)" awarded to Anshul Verma at Banaras Hindu University, Varanasi.

REFERENCES

1. Sharun, K., & Dhama, K. (2021). India's role in COVID-19 vaccine diplomacy. *Journal of Travel Medicine*, 28(7), 1–4. https://academic.oup.com/jtm/article/28/7/taab064/6231165

2. Gupta, M., Goel, A. D., & Bhardwaj, P. (2021). The COWIN portal – current update, personal experience and future possibilities. *Indian Journal of Community Health*, 33(2), 414. https://scholar.archive.org/work/mdn3nh5tqzgdvgxebblranxewq/access/wayback/https://www.iapsmupuk.org/journal/index.php/IJCH/article/download/2208/1200

3. Masse, M. (2011). *REST API Design Rulebook*. O'Reilly Media, Inc. https://books.google.co.in/books?hl=en&lr=&id=eABpzyTcJNIC&oi=fnd&pg=PR3&dq=Masse,+M.+(2011).+REST+API+Design+Rulebook.+O%E2%80%99Reilly+Media,+Inc+.&ots=vAUD-3ibHC&sig=I65Oz64oFAdAt-SMB-Yh6_j5eWg&redir_esc=y#v=onepage&q&f=false

4. Corporation, D. I., Division, N. e.-G., & Government of India. (2021). *API Setu: Co-WIN Public APIs*. Retrieved from API Setu: https://apisetu.gov.in/public/marketplace/api/cowin.

5. Marsicano, K., Gardner, B., Phillips, B., & Stewart, C. (2019). *Android Programming: The Big Nerd Ranch Guide*. Big Nerd Ranch Guides. https://api.pageplace.de/preview/DT0400.9780135218136_A41216470/preview-9780135218136_A41216470.pdf

6. Aditya, S. K., & Karn, V. K. (2014). *Android SQLite Essentials*. Packt Publishing. http://www.bibliotecasyarchivos.net/sitio_bdusam/bibliotecadigital/ebook/facultad_de_ciencias_empresariales/Licenciatura_en_computacion/Aplicaciones_Moviles/Android_SQLite_Essentials_nodrm.pdf

7. Blundell, P., & Milano, D. T. (2015). *Learning Android Application Testing*. Packt Publishing Limited. https://books.google.com/books?hl=en&lr=&id=TP2rBwAAQBAJ&oi=fnd&pg=PP1&dq=Blundell,+P.,+%26+Milano,+D.+T.+(2015).+Learning+Android+Application+Testing.+Packt+Publishing+Limited.&ots=1CIDZk-0_x&sig=Gts86VC_B5LGlsl8r2fB8Papxao

8. Vashisht, T. (2021). *VaccinateMe | Get Notified When Vaccinations Are Available*. Retrieved from HealthifyMe: https://apps.healthifyme.com/vaccinateme/.

9. Azar, Shyam, Anurag, & Akshay. (2021). *Vaccine Slot Notifier*. Retrieved from Getjab: https://getjab.in/.

10. Khan, I. (2021). *CoWIN Vaccine Tracker*. Retrieved from Google Play: https://play.google.com/store/apps/details?id=com.imranapps.vaccinetracker.

11. Dhameliya, D. (2021). *Vaccine Slot Notifier*. Retrieved from Google Play: https://play.google.com/store/apps/details?id=com.darshit.vaccinenotifier.

12. Bytes, P. (2021). *Vaccine Slot Tracker*. Retrieved from Google Play: https://play.google.com/store/apps/details?id=com.pepperbytes.vaccineslottracker.

13. Sharma, S. (2021). *Covid-19 Vaccination Slot Finder for 18 Years and above in India*. Retrieved from Find Slot: https://findslot.in/.

14. Saha, S. (2021). *SRvSaha/CoWin Vaccine Slot Finder*. Retrieved from GitHub: https://github.com/SRvSaha/CoWinVaccineSlotFinder.

15. Kumar, A., Verma, A. K., & Barik, M. (n.d.). Aarogya Setu App (APA) for Covid-19: Recent advances and future prospective. *A Text Book of the SARS-CoV-2*, pp. 179–186. Mahi Publication. https://www.mahipublication.com/bookdetails/a-text-book-of-the-sars-cov-2-guidelines-and-protocol-development/OTY=/

16. Portal-MoHFW, N. H. (2021). *Co-WIN Vaccinator App.* Retrieved from Google Play Store: https://play.google.com/store/apps/details?id=com.cowinapp.app.
17. Jalote, P. (2003). *An Integrated Approach towards Software.* Narosa Publishing House.
18. Ramez, E., & Shamkant, N. (2017). *Fundamentals of Database System.* Pearson Education.
19. Torlak, E. (2015). *UML Sequence Diagrams.* Retrieved from University of Washington Computer Science & Engineering: https://courses.cs.washington.edu/courses/cse403/15sp/lectures/L10.pdf.
20. Friesen, J. (2010). *Learn Java for Android Development.* Apress.
21. Raynaldo, C. (2012). *Android UI Design with Xml: Tutorial Book.* Createspace Independent Pub.
22. Sanderson, P., Hipp, R., Shavers, B., Mahalik, H., & Zimmerman, E. (2018). *SQLite Forensics.*Independently published. https://dl.acm.org/doi/abs/10.5555/3265036
23. Documentation I Android Developers. (n.d.). Retrieved 2021, from Android Developers: https://developer.android.com/docs.
24. Wharton, J. (2010). *Retrofit.* Retrieved from Retrofit: https://square.github.io/retrofit/.
25. Mao, K., Harman, M., & Jia, Y. (2017). Robotic testing of mobile apps for truly black-box automation. *IEEE Software,* 34(2), 11–16.
26. Gunasekera, S. (2012). *Android Apps Security.* Berkeley, CA: Apress. https://link.springer.com/content/pdf/10.1007/978-1-4842-1682-8.pdf

17 Mathematical Modeling of Glucocerebrosidase Signalling Networks Linked to Neurodegeneration

*Hemalatha Sasidharakurup, Dheeraj Pisharody, Haritha Bose, Dijith Rajan, and Shyam Diwakar**

Amrita Mind Brain Center, Amrita Vishwa Vidyapeetham, Amritapuri campus

School of Biotechnology, Amrita Vishwa Vidyapeetham, Amritapuri campus

CONTENTS

* shyam@amrita.edu

DOI: 10.1201/9781003320333-17

17.1   INTRODUCTION

Human brain is a complex system with large signaling networks of interconnected neurons that control emotions, motor skills, thoughts, memory, and other body functions [1]. Recent advances in computational and systems biology help researchers to understand how individual neurons are linked to these networks to collectively activate different parts of the brain responsible for different functions [2]. Recent developments in mapping and recording techniques help scientists to analyze the distributed neuronal connections from single cell to whole brain network, bridging laboratory measurable and immeasurable parameters [3]. In the context of brain related disorders, analyses on synaptic malfunction of a single neuron due to intracellular changes alone cannot help to identify how it could alter any particular behavior of the entire system. However, it is possible if there is a network of many interconnected neurons, as changes in any single signaling process could show how it affects different regions of the brain. Laboratory experiments have many limitations to measure cellular level changes in neurons and connect it to the whole-brain behavior. One of the challenges is to find out how changes in cellular parameters (genes/proteins) are responsible for emergent properties that cause neurodegeneration in major degenerative diseases including Alzheimer's (AD) and Parkinson's (PD). At the systems level, modeling of intracellular biochemical reactions, interconnections, and cross-talk within the signaling networks provides a better understanding of how a single gene or protein can communicate the entire system causing these disease conditions. This may help clinicians to find better biomarkers for early detection and to delay disease progression as there's no specific treatment for such diseases.

This paper contributes such a systems-level mathematical model based on biochemical systems theory (BST) and ordinary differential equations (ODE) to show how mutations in glucocerebrosidase (GBA) gene can affect the brain networks from cellular level leading to neurodegeneration as observed in AD/PD. BST-based models can represent intracellular reactions in mathematical equations by applying chemical kinetics, ODEs, initial concentrations, and rate laws, where time-series data can show how concentration fluctuations in every signaling molecule cause changes in the entire pathway network. AD and PD are most common age-dependent neurodegenerative diseases where there is no medication to completely stop the diseases. The GBA gene is an essential protein coding gene that helps in the production of lysosomal enzyme called β-glucocerebrosidase (GCase). Studies reported many risk factors including genetic causes, chemical toxins, and other mutagens can cause mutations in GBA gene that disturbs autophagy–lysosomal pathways, a cellular degradation system for intracellular macromolecules [4]. Recent studies have shown that damage in these pathways along with GBA mutations can trigger other signaling pathways leading to neurodegeneration, slowly developing AD and PD [5–7]. Studies have also reported that protein accumulation can cause cellular stress responses and mitochondrial impairment, which may result in the accumulation of α-synuclein, another protein involved in the PD condition [4]. In our previous AD and PD models, we have already explained how disturbances in the autophagy lysosomal pathway could affect other parts of the cellular network to initiate neurodegeneration [8]. Since GBA has an important role in autophagy–lysosomal pathway that leads to neurodegeneration, modeling its cellular networks would help to find important biomarkers related to AD and PD. This model could help clinicians in early detection of the disease and treatment or to identify better therapeutic targets to stop disease progression. In this model, we have mathematically reconstructed biochemical reactions, bifurcations, interconnections, cross-talks, and feedback loops between signaling molecules for a better understanding of shared mechanisms and commonness of GBA pathway involved in the autophagy–lysosomal dysfunction in both AD and PD. Apart from the contributions of recent computational models that have explained the role of autophagy-lysosomal pathways in neurodegeneration [9], our study provides a detailed systems-level dynamic model based on laboratory experimental data that predict common biomarkers for both AD and PD.

17.2 METHODS

In this model, the biochemical reactions in the GBA gene pathway network related to autophagy–lysosomal dysfunctions have been reconstructed. All the biochemical reactions, their initial concentrations, and reaction rates in the modelled pathways were collected from experimental studies through literature survey. Protein/gene species, other signaling biomolecules, and associated reactions were drawn using the pathway modeling tool, CellDesigner (www.celldesigner.org) [10] (see Figure 17.1). Genes, proteins, and other biomolecules were represented as nodes with unique shapes in the pathway networks. Each molecular process was represented by connecting the nodes using specific reaction icons that can be selected from the toolbar. Proteins inside the human body were detected and quantified using biological laboratory experimental techniques including western blotting, fluorescent assays, radio-enzymatic assays, and ELISA. Literature survey was done to include laboratory studies that provided initial concentration values of proteins from these experiments in both diseased state and normal state (see Tables 17.1 and 17.2). These initial values were assigned to every biomolecule in the simulation tab. CellDesigner generated kinetic laws based on the diagrammatic representation of every reaction in the pathway. In this model, law of mass action, Michaelis Menten kinetics, hill equations, convenience kinetics etc., were used as described in our previous models [11].

Law of mass action stated that the rate of a reaction is proportional to the product of the concentrations of reactants. Consider a reaction of dopamine synthesis from tyrosine and L-dopa with the help of tyrosine hydroxylase (TH).

$$
\begin{array}{c}
\text{TH} \\
\downarrow \\
\text{Tyrosine} \longrightarrow \text{L-Dopa} \longrightarrow \text{Dopamine}
\end{array}
$$

According to mass actions, V1, the rate of L-Dopa is equal to the concentration of tyrosine and TH multiplied with the rate constant.

$$\text{i.e, L-dopa} = k1[\text{Tyrosine}][\text{TH}] \tag{17.1}$$

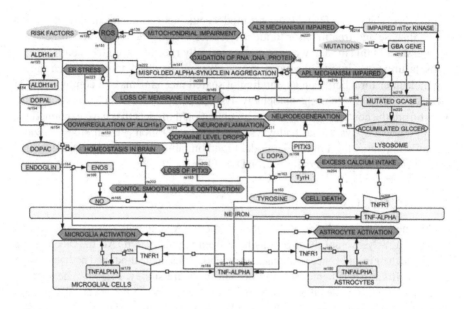

FIGURE 17.1 Cellular mechanisms affected by GBA mutations leading to Parkinson's condition.

Similarly, the rate of dopamine V2 can be represented as,

$$\text{Dopamine} = k2[\text{L-Dopa}] \tag{17.2}$$

where k1 and k2 are reaction rate constants. Hence, rate of change of L-Dopa can be represented as,

$$d/dt[\text{L-dopa}] = V1 - V2 \tag{17.3}$$

Michaelis-Menten's kinetics included both reversible and irreversible reactions.

$$v = \frac{d[p]}{dt} = \frac{v_{\max}[s]}{k_m + [s]} \tag{17.4}$$

where $v_{\max}$ is the maximum rate achieved by the system, at saturating substrate concentration in relation of reaction rate v to $[S]$; $[S]$ is the concentration of a substrate S. k_m is Michealis constant is the rate constant. And the rate constant k_m (Michaelis constant) is equal to $[S]$, when V is half of $v_{\max}$. It relates velocity of the reaction to the concentration of substrates. If V is the rate of reaction, it changes with the substrate concentration for different enzymes.

The tool used ODE solvers to solve ODEs and their multiple derivatives generally computed by Euler's or Runge–Kutta methods. By providing initial concentrations and rates for every species node, the concentration versus time graphs were plotted from the simulations. By calculating concentration variations as a function of time, dynamical interactions among network nodes were analyzed according to varying parameters of the disease. This helped to identify mechanisms, predict behavior, and possible biomarkers for the disease progression. The discrepancies between species regulations and their behavior in both diseased and normal states in this model can be used to provide new insights on the early onset of AD/ PD, which can contribute to diagnostics anticipated forthcoming developments.

17.3   BIOCHEMICAL PATHWAY MODELING

17.3.1   MODELING MUTATED GBA-RELATED CELLULAR NETWORKS IN AD CONDITION

The GBA gene was an essential protein coding gene helping in the production of lysosomal enzyme called β-glucocerebrosidase (GCase). GCase degrades glucocerebroside (a cellular membrane content) into glucose and ceramide to prevent accumulation of glucosylceramide (GlcCer). In AD and PD, low GBA and GCase cause a progressive increase in GlcCer concentrations. This resulted in lack of sphingolipid metabolism, which was essential for many biological functions and membrane integrity. Mutations in GBA genes could be a significant factor in causing PD because it could lead to the reduced activity of the enzyme encoded by GBA genes, GCase [12]. From here, conditions worsened as the GlcCer accumulated inside the lysosomes, which could cause the lack of sphingolipid metabolism that led to lack of membrane integrity. As normal functions of neurons including neuronal connectivity and synaptic transmission relied on plasma membranes, this could cause stress in the cell, triggering the release of reactive oxygen species (ROS), which could have caused mitochondrial impairment, and to release more ROS [13]. ROS could oxidase RNA, DNA, or lipids present in the cells, which may have led to the weakening of the plasma membrane integrity in a severe way, and due to this, autophagy lysosomal mechanism that cleans up the interior of the cell gets impaired.

The reactions were reconstructed showing decrease in GCase activity led to accumulation of GlcCer, which may cause loss of cell membrane integrity and autophagy–lysosomal dysfunction [4]. GlcCer-inhibited ApoE, a transporter protein in which unsaturated fatty acids were transported, also helped in the clearance of β-amyloid produced by the microglial cells. β-amyloid was one of the major proteins involved in the progression of AD. They aggregated to form amyloid plaques (insoluble protein aggregates) between neurons causing neurodegeneration that affected memory and other cognitive functions in AD [14]. Due to loss of membrane integrity, ApoE needed the membrane receptor serotonin to integrate into the cytoplasmic region. This promoted β-amyloid accumulation that resulted in neuroinflammation, another condition leading to neurodegeneration. ApoE3 and ApoE4 both helped in β-amyloid clearance by acting as a ligand to the cellular receptor, sortilin. These apolipoproteins carried the plaque-forming β-amyloid and bound to the receptor. The receptors took the apolipoproteins with β-amyloid and carried them to the lysosomal compartment in the cell. Under normal conditions, after the degradation, recycling of the receptor to the surface will take place. But APOE4 acted as a risk factor for AD, by disrupting the recycling mechanism of the surface receptors. A low level of APOE was transported to the inner cell, which resulted in the aggregation of β-amyloid concentration that enhanced the β-amyloid production. APOE4 also led to hyper phosphorylation of tau protein, which formed tangles and enhanced neurodegeneration in AD condition. These proteins were also associated with the impairment of autophagy–lysosomal mechanism, which led to neuroinflammation and neurodegeneration. They initiated the production of high levels of P-IRS1(phospho-insulin receptor substrate-1), which disturbed insulin signaling that may cause insulin resistance. Eventually, all this together affects different regions of the brain, which begins to shrink. The first symptoms of this disease may be the problems with vision or language rather than memory.

17.3.2 MODELING OF MUTATED GBA CELLULAR NETWORKS IN PD CONDITION

α-synuclein is a major protein encoded by SNCA gene in which its abnormal aggregation plays an important role in the pathogenesis of PD. Membrane integrity was required for the proper function of autophagy–lysosomal mechanism. Due to loss of membrane integrity, impaired the autophagy–lysosomal mechanism became a perfect condition for the α-synuclein aggregation, and there was a double negative feedback relation between the lysosomal mechanism and aggregated α-synuclein [4]. Recent studies have shown that the association between GBA gene mutation and α-synuclein aggregation is inversely related; GCase activity decreased when α-synuclein aggregation increased and vice versa [4]. The autophagic–lysosome reformation mechanism got disturbed as GCase indirectly controlled mTOR, which regenerated functional lysosomes. Studies have shown significant decrease of Phospho-S6K in PD patients with GBA mutations, which is a marker of mTOR activity [15,16]. PITX3 (paired-like homeodomain transcription factor 3), an important transcriptional factor that regulated the enzyme TH, which catalyzed the formation of dopamine from L-dopa and ALDH1A1 (Aldehyde Dehydrogenase 1 Family Member A1) genes. Loss of PITX3 downregulated ALDH1A1 enzymes and upregulated DOPAL. DOPAL was an important precursor metabolite of dopamine. Upregulated DOPAL triggered α-synuclein oligomerization [17,18]. All these reactions together led to increased accumulation of α-synuclein that activated neuroinflammation. α-synuclein also activated the astrocytes, which released TNFα and other pro-inflammatory cytokines [19]. This TNFα induces the activation of microglial cells (immune cells of the brain) to produce more TNFα. Previous studies have shown that TNFα is a down regulator of a cell surface protein endoglin that distract the nitric oxide pathway, which can cause abnormalities in movement [20].

17.3.3 Accumulation of GlcCer, Low Levels of GCase, Glucose, and Ceramide in AD Compared to Normal Condition

The concentration differences of protein levels were analyzed in AD control and diseased conditions. In AD normal state (see Figure 17.2a), simulations have shown increased levels of glucose and ceramide production. There was a slight increase in APOE concentration that slowly reached a stable level. APOE may influence increased formation of β-amyloid plaques in the brain, which promotes the risk of AD. Simulations show clearance of β-amyloid compared to diseased state. There was a gradual decrease in the concentration levels of GlcCer with time. This was because accumulation was prevented by GCase activity. Overall, the results indicated that

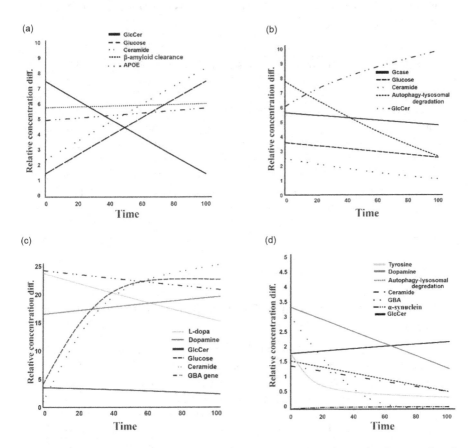

FIGURE 17.2 Relative concentration differences of genes and proteins in control and diseased conditions. (a) Low levels of GlcCer accumulation and high levels of glucose and ceramide in normal condition. (b) High levels of GlcCer accumulation, low levels of GCase, glucose, and ceramide in diseased AD condition. (c) High levels of dopamine, ceramide, glucose, GBA, and low GlcCer accumulation in normal condition compared to PD. (d) Low dopamine, tyrosine, GBA, and high accumulation of GlcCer in PD condition.

in normal condition, GBA function maintained the homeostasis of the system. In diseased condition (see Figure 17.2b), simulations have shown a gradual decrease in the autophagic lysosomal degradation that led to the accumulation of amyloid plaques and neurodegeneration, which is linked to memory loss and other neurocognitive problems in AD [14]. There was an elevation in the levels of GlcCer accumulation compared to normal conditions. The levels of GCase, glucose, and ceramide were decreasing compared to normal conditions. This is because mutated GBA gene cannot produce enough GCase to break down glucocerebroside into glucose and ceramide.

TABLE 17.1
Parameter Estimation of Initial Concentrations in AD Conditions from Published Literature

| Gene/Protein | Concentration | | Experimental Model | Ref. |
	Control	Diseased		
Amyloid β	1.7 ± 2.3 mg	6.5 ± 5.0 mg	Human	[21]
TNFα	8.1 pg/mL	2.5 ± 1.25 pg/mL	Human	[22]
Plasma tau	4.43 ± 2.83 pg/mL	8.80 ± 10.1 pg/mL	Human	[23]
MCP-1	66 ± 50 pg/mL	43 ± 28 pg/mL	Mice	[24]
IL-10	1.06 pg/mL	1.3–41.7 pg/mL	Human	[25]
APOE – 4	46.6 μg/mL	40 ng/mL	Human	[26]
APOE	40 ng/mL	46.1 ± 19.0 μg/mL	Human	[27]
ROS	12.7 ± 0.17 pmol/min/mg	32.0 ± 2.2 pmol/min/mg	Guinea pigs	[28]
Calcium	7.2 ± 2.6 μg/mL	8.1 ± 2.1 μg/mL	Human	[29]

TABLE 17.2
Parameter Estimation of Initial Concentrations for PD from Published Literature

| Gene/Protein | Concentration | | Experimental Model | Ref. |
	Control	Diseased		
GlcCer	7.4 μmol/L	16.3 μmol/L	Human	[30]
TNFα	110.7 ± 87.2 pg/mL	196.9 ± 121.2 pg/mL	Human	[31]
Mutated GCase	-	0.22–1.04 mU/mL	Human	[32]
α-synuclein	0.42 pg/mL	1.29 pg/mL	Human	[33]
DOPAC	3.48 ± 0.66	0.164 ± 0.057	Human	[34]
GBA	55.1 ± 8.5	50.5 ± 6.6	Human	[35]
DOPAL	0.496 ± 0.164	0.071 ± 0.036	Mice	[36]
Microglia	0.5%–16.6%	0.5%–16.6%	Mice	[37]
Astrocytes	20%–40%	20%–40%	Human	[38]

17.3.4 High Levels of GlcCer, β-amyloid, and Low GBA in Diseased State

In the PD control (see Figure 17.2c), the level of L-dopa was decreasing with the increase in production of dopamine. GBA was gradually decreasing, while the production of glucose and ceramide was constantly increasing. The levels of GlcCer were decreasing with time compared to diseased conditions. There was no neuro-inflammation observed in normal conditions. In diseased state (see Figure 17.2d), there was a sudden drop in the enzyme TH, and due to this, dopamine levels were gradually decreasing. TH helps in the production of dopamine. Simulations show low levels of GBA, glucose, and ceramide concentrations with time. The autophagy–lysosomal degradation was low. This led to high accumulation of GlcCer and α-synuclein. Low level of dopamine is linked to tremor, one of the onset symptoms of PD. Results indicated that low concentration of GBA and GCase led to gradual increase in GlcCer concentration levels, which affected sphingolipid metabolism causing loss of membrane integrity. As α-synuclein concentration increased, microglial cells were activated and triggered the release of various pro-inflammatory cytokines. This may activate astrocytes and again release proinflammatory cytokines including TNFα, creating a feedback loop. This can also lead to neuroinflammation and neurodegeneration. The GBA gene has also been linked to severe conditions such as cognitive decline or depression [39].

17.4 DISCUSSION

The objective of this study was to model and analyze the involvement of GBA genes, which cause neurodegeneration in AD and PD. Although mutations in GBA genes are quite correlated with Gaucher's disease, studies also have reported its role in the pathogenesis of many other diseases such as Parkinson's, Alzheimer's, dementia with Lewy bodies and sleep behavior disorders [40]. In this study, by reconstructing the GBA-related cellular network connections, we could analyze the biochemical reactions, cross-talks, and feedback mechanisms that can be affected due to GBA mutations associated with AD and PD. The results from the model were validated against previous laboratory experimental studies and mapped with clinical symptoms.

The model showed that in both the diseased conditions, mutations in the GBA gene also cause changes in GCase activity and GlcCer accumulation disturbing the autophagy–lysosomal pathway. In the case of PD, as GCase activity decreases, α-synuclein aggregation increased. The model suggests that since accumulated GlcCer reduce cell membrane rigidity, it affected lysosomal enzyme activity and α-synuclein degradation resulting in impaired autophagy–lysosomal mechanism and accumulated α-synuclein [41]. Abnormal increase in α-synuclein levels could activate astrocytes and microglial cells to release proinflammatory cytokines including TNFα, resulting in neurodegeneration. In AD, since β-amyloid deposition is a lysosomal entity, mutations in GBA genes could inhibit the autophagy–lysosomal mechanism that clears β-amyloid aggregation. The model suggests that as GCase level decreases, it led to the accumulation of GlcCer and resulted in loss of membrane integrity and dysfunction of the autophagy–lysosomal system. Finally, this triggered neuroinflammation and neurodegeneration affecting many parts of the brain related to memory and cognitive functions [42].

According to this model, decrease of GBA gene influenced autophagy–lysosomal dysfunction, protein aggregation, ROS formation, and neuroinflammation in both AD and PD networks. Along with low GBA levels, these factors also can be considered as better biomarkers for the disease progression. Cognitive deterioration and depressive symptoms can be linked to mutations of GBA gene. This model helps in tracking changes due to GBA mutations in AD/PD conditions to understand how these systemic changes affect different parts of the system. Although the model makes predictions based on previous data, experimentation may be needed to map the predictions to actual changes in animal models going through in vitro and in vivo conditions. However, at this stage, these biomarkers may help clinicians to connect model predictions with laboratory indices and symptoms to relate disease progression over the course of time.

17.5 CONCLUSION

The complex biochemical networks that have been modelled in this study helped to analyze and understand how GBA mutations cause neurodegeneration in AD and PD. The model also helped to understand common pathways activated in both AD and PD due to GBA mutations. This model can be used to find different therapeutic strategies and targets that could be aimed at bringing a novel biomarker or an all-time cure for such diseases.

ACKNOWLEDGMENTS

This work derives direction and ideas from the Chancellor of Amrita University, Sri Mata Amritanandamayi Devi. This study has been partially supported by the Department of Science and Technology Grant DST/CSRI/2017/31, Government of India and Embracing the World research for a cause Initiative.

REFERENCES

1. Bassett DS, Gazzaniga MS (2011) Understanding complexity in the human brain. *Trends Cogn Sci* 15:200–209. https://doi.org/10.1016/j.tics.2011.03.006.
2. Siegelmann HT (2010) Complex systems science and brain dynamics. *Front Comput Neurosci* 4:7. https://doi.org/10.3389/fncom..2010.00007.
3. Sporns O (2016) Connectome networks: From cells to systems. In: Kennedy H, Van Essen DC, Christen Y (eds), *Research and Perspectives in Neurosciences*. Switzerland: Springer Verlag, pp. 107–127.
4. Do J, McKinney C, Sharma P, Sidransky E (2019) Glucocerebrosidase and its relevance to Parkinson disease. *Mol Neurodegener* 14:1–16.
5. Riboldi GM, Di Fonzo AB (2019) GBA, Gaucher disease, and Parkinson's disease: From genetic to clinic to new therapeutic approaches. *Cells* 8:364. https://doi.org/10.3390/cells8040364.
6. Brockmann K (2020) GBA-associated synucleinopathies: Prime candidates for alpha-synuclein targeting compounds. *Front Cell Dev Biol* 8:1–7.

7. Avenali M, Blandini F, Cerri S (2020) Glucocerebrosidase defects as a major risk factor for Parkinson's disease. *Front Aging Neurosci* 12:97.
8. Sasidharakurup H, Nair L, Bhaskar K, Diwakar S (2020) Computational Modelling of TNFα Pathway in Parkinson's Disease – A Systemic Perspective. In: Cherifi, H., Gaito, S., Mendes, J., Moro, E., Rocha, L. (eds) Complex Networks and Their Applications VIII. COMPLEX NETWORKS 2019. Studies in Computational Intelligence, vol. 882, pp. 762–773. Springer, Cham. https://doi.org/10.1007/978-3-030-36683-4_61.
9. Han K, Kim SH, Choi M (2020) Computational modeling of the effects of autophagy on amyloid-β peptide levels. *Theor Biol Med Model* 17:1–16.
10. Funahashi A, Matsuoka Y, Jouraku A, et al (2006) Celldesigner: A modeling tool for biochemical networks. In: *Proceedings of the 38th Conference on Winter Simulation, Monterey, CA.* pp. 1707–1712.
11. Sasidharakurup H, Diwakar S (2020) Computational modelling of TNFα related pathways regulated by neuroinflammation, oxidative stress and insulin resistance in neurodegeneration. *Appl Netw Sci* 5.
12. Gegg ME, Schapira AHV (2018) The role of glucocerebrosidase in Parkinson disease pathogenesis. *FEBS J* 285:3591–3603.
13. van Echten-Deckert G, Herget T (2006) Sphingolipid metabolism in neural cells. *Biochim Biophys Acta - Biomembr* 1758:1978–1994.
14. Guillozet AL, Weintraub S, Mash DC, Marsel Mesulam M (2003) Neurofibrillary tangles, amyloid, and memory in aging and mild cognitive impairment. *Arch Neurol* 60:729–736.
15. Chen Y, Yu L (2017) Recent progress in autophagic lysosome reformation. *Traffic* 18:358–361.
16. Magalhaes J, Gegg ME, Migdalska-Richards A, et al (2016) Autophagic lysosome reformation dysfunction in glucocerebrosidase deficient cells: Relevance to Parkinson disease. *Hum Mol Genet* 25:3432–3445. https://doi.org/10.1093/hmg/ddw185.
17. Masato A, Plotegher N, Boassa D, Bubacco L (2019) Impaired dopamine metabolism in Parkinson's disease pathogenesis. *Mol Neurodegener* 14:1–21.
18. Bruggeman FJ, Bakker BM, Hornberg JJ, Westerhoff H. V (2006) Introduction to computational models of biochemical reaction networks. In: Kriete, A, Eils, R (eds) *Computational Systems Biology.* London: Elsevier Inc. pp. 127–148.
19. Lee HJ, Kim C, Lee SJ (2010) Alpha-synuclein stimulation of astrocytes: Potential role for neuroinflammation and neuroprotection. *Oxid Med Cell Longev* 3:283–287.
20. Dallas NA, Samuel S, Xia L, et al (2008) Endoglin (CD105): A marker of tumor vasculature and potential target for therapy. *Clin Cancer Res* 14:1931–1937.
21. Roberts BR, Lind M, Wagen AZ, et al (2017) Biochemically-defined pools of amyloid-β in sporadic Alzheimer's disease: Correlation with amyloid PET. *Brain* 140:1486–1498. https://doi.org/10.1093/brain/awx057.
22. Cacabelos R, Alvarez XA, Franco-Maside A, et al (1994) Serum tumor necrosis factor (TNF) in Alzheimer's disease and multi- infarct dementia. *Methods Find Exp Clin Pharmacol* 16:29–35.
23. Yang SY, Chiu MJ, Chen TF, et al (2017) Analytical performance of reagent for assaying tau protein in human plasma and feasibility study screening neurodegenerative diseases. *Sci Rep* 7:1–12. https://doi.org/10.1038/s41598-017-09009-3.
24. Porcellini E, Ianni M, Carbone I, et al (2013) Monocyte chemoattractant protein-1 promoter polymorphism and plasma levels in alzheimer's disease. *Immun Ageing* 10:6. https://doi.org/10.1186/1742-4933-10-6.
25. Csuka E, Morganti-Kossmann MC, Lenzlinger PM, et al (1999) IL-10 levels in cerebrospinal fluid and serum of patients with severe traumatic brain injury: Relationship to IL-6, TNF-α, TGF-β1 and blood-brain barrier function. *J Neuroimmunol* 101:211–221. https://doi.org/10.1016/S0165-5728(99)00148-4.

26. Simon R, Girod M, Fonbonne C, et al (2012) Total ApoE and ApoE4 isoform assays in an Alzheimer's disease case-control study by targeted mass spectrometry (n = 669): A pilot assay for methionine-containing proteotypic peptides. *Mol Cell Proteomics* 11:1389–1403. https://doi.org/10.1074/mcp.M112.018861.

27. Rezeli M, Zetterberg H, Blennow K, et al (2015) Quantification of total apolipoprotein E and its specific isoforms in cerebrospinal fluid and blood in Alzheimer's disease and other neurodegenerative diseases. *EuPA Open Proteomics* 8:137–143. https://doi.org/10.1016/j.euprot.2015.07.012.

28. Tretter L, Adam-Vizi V (2004) Generation of reactive oxygen species in the reaction catalyzed by α-ketoglutarate dehydrogenase. *J Neurosci* 24:7771–7778. https://doi.org/10.1523/JNEUROSCI.1842-04.2004.

29. Paglia G, Miedico O, Cristofano A, et al (2016) Distinctive pattern of serum elements during the progression of Alzheimer's disease. *Sci Rep* 6:1–12. https://doi.org/10.1038/srep22769.

30. Gomes Muller MV, Petry A, Pinheiro Vianna L, et al (2010) Quantification of glucosylceramide in plasma of Gaucher disease patients. *Brazilian Journal of Pharmaceutical Sciences*, 46:643–649

31. Li G, Wu W, Zhang X, et al (2018) Serum levels of tumor necrosis factor alpha in patients with IgA nephropathy are closely associated with disease severity. *BMC Nephrol* 19:1–9. https://doi.org/10.1186/s12882-018-1069-0.

32. Oftedal L, Maple-Grødem J, Førland MGG, et al (2020) Validation and assessment of preanalytical factors of a fluorometric in vitro assay for glucocerebrosidase activity in human cerebrospinal fluid. *Sci Rep* 10:1–8. https://doi.org/10.1038/s41598-020-79104-5.

33. Lin CH, Yang SY, Horng HE, et al (2017) Plasma α-synuclein predicts cognitive decline in Parkinson's disease. *J Neurol Neurosurg Psychiatry* 88:818–824. https://doi.org/10.1136/jnnp-2016-314857.

34. Blandini F, Fancellu R, Martignoni E, et al (2001) Plasma homocysteine and l-DOPA metabolism in patients with Parkinson disease. *Clin Chem* 47:1102–1104. https://doi.org/10.1093/clinchem/47.6.1102.

35. Gegg ME, Sweet L, Wang BH, et al (2015) No evidence for substrate accumulation in Parkinson brains with GBA mutations. *Mov Disord* 30:1085–1089. https://doi.org/10.1002/mds.26278.

36. Goldstein DS, Sullivan P, Holmes C, et al (2013) Determinants of buildup of the toxic dopamine metabolite DOPAL in Parkinson's disease. *J Neurochem* 126:591–603. https://doi.org/10.1111/jnc.12345.

37. Bachiller S, Jiménez-Ferrer I, Paulus A, et al (2018) Microglia in neurological diseases: A road map to brain-disease dependent-inflammatory response. *Front Cell Neurosci* 12:488.

38. Sloan SA, Barres BA (2014) Mechanisms of astrocyte development and their contributions to neurodevelopmental disorders. *Curr Opin Neurobiol* 27:75–81.

39. Kasten M, Marras C, Klein C (2017) Nonmotor signs in genetic forms of Parkinson's disease. In: Ray Chaudhuri, K, Tolosa, E, Schapira, AHV, Poewe, W (eds), *International Review of Neurobiology*. Oxford: Academic Press Inc.. pp. 129–178.

40. Eblan MJ, Nguyen J, Ziegler SG, et al (2006) Glucocerebrosidase mutations are also found in subjects with early-onset Parkinsonism from Venezuela. *Mov Disord* 21:282–283.

41. Stojkovska I, Krainc D, Mazzulli JR (2018) Molecular mechanisms of α-synuclein and GBA1 in Parkinson's disease. *Cell Tissue Res* 373:51–60.

42. Llorente-Ovejero A, Martínez-Gardeazabal J, Moreno-Rodríguez M, et al (2021) Specific phospholipid modulation by muscarinic signaling in a rat lesion model of Alzheimer's disease. *ACS Chem Neurosci* 12:2167–2181.

18 Empirical Analysis of The Performance of Routing Protocols in Opportunistic Networks

Mohini Singh and Anshul Verma
Banaras Hindu University

Pradeepika Verma
Indian Institute of Technology Patna

CONTENTS

18.1 INTRODUCTION

Opportunistic Networks can be seen as an extended version of Mobile Ad hoc Networks (MANETs) [1–3]. Due to this, they derive challenges of Ad hoc Networks along with some new ones. Some important characteristics of Opportunistic Networks are: (a) The end-to-end path from sender to receiver hardly exists because a network is partitioned into smaller and independent regions which are located separately.

DOI: 10.1201/9781003320333-18

(b) Contacts between nodes in a network are intermittent; in other words, the probability of contacts between nodes is very low. (c) Both reconnections and disconnections occur frequently as the nodes are mobile and/or in order to save power they may turn it off. (d) Nodes that are in between the source and destination follow the store-carry-and-forward paradigm for message transmission. (e) Performance of any established link varies significantly. Opportunistic Networks are similar to the Delay-Tolerant Networks (DTN) [4–5] except that the point of connection/disconnection or timing of connection/disconnection is not known in advance in Opportunistic Networks whereas DTN is a network of connected clusters of nodes in which nodes are well connected within a cluster. However, connections between clusters are intermittent but points of connections between clusters and sometimes the timing of connections are already known. Nodes in Opportunistic Networks are generally mobile like vehicles or pedestrian users but they can be fixed also. Discovery and communication among nodes can occur by using any or all kinds of communication media. Opportunistic Networks may start operations as a single node initially, which is called a seed node, and then it may expand itself by adding some helper nodes (foreign nodes). Applications of Opportunistic Networks are in the cases of preparedness for emergency and response activities.

Long propagation delays and other various disruptions in Opportunistic Networks make it unsuitable for the use of traditional routing schemes of MANETs and many other Internet protocols like TCP/IP as they may fail to deliver performance and delivery of messages in Opportunistic Networks. Due to the absence or uncertainty of end-to-end path from source to destination nodes, Opportunistic Networks use mobility of nodes and method of local forwarding to transfer the data. Intermediate nodes participate just like a relay node to forward the data, and the combined effect causes delivery of the data to the destination node. Opportunistic Networks are of intermittent nature; hence, intermediate nodes' encountering other nodes is a matter of probability and it doesn't occur frequently or consistently. It might happen that there is no suitable intermediate node that is promisingly taking the data closer to the destination. In such situations, the data has to be directly transferred to the destination node as soon as direct contact arises. In order to handle such situations, nodes may need to store the packets for a longer duration of time. Intermediate nodes apply the technique of store-carry-and-forward and store data till there is a forwarding opportunity available towards the destination.

The store-carry-and-forward method increases the chances of successful delivery of data significantly to the destination but inherently introduces a larger delay as the storage of data is taking place at intermediate nodes and there is a waiting period for connection availability towards the destination. Nodes must have buffer storage so that all messages can be stored for spans of time that are of unpredictable duration, i.e. till the next contact is established. Hence, the requirement for storage space is directly linked to the number of messages transferred through the network. The task of routing and forwarding messages in Opportunistic Networks is challenging since there lies is an uncertainty regarding the intermittent behaviour and mobility of nodes. Research works in Opportunistic Networks mostly revolve around routing and forwarding as finding successful routes to the intended destination is one of the most tedious tasks. Hence, a need to design new protocols always arises in Opportunistic

Networks which try for maximising the data delivery rate and minimising delays in transmission towards the destination.

18.2   ROUTING PROTOCOLS

Routing protocols [6–8] are a strategy or procedure for transferring data packets among the nodes in the network. Data packets can be transferred in the network by blind or controlled flooding or with the help of encounter history of nodes to get the best path. Therefore, routing protocols can be categorised as flooding or probability based. On the basis of the number of copies of a message forwarded in the network, it can also be categorised into single or multiple copy-based schemes. There are some issues in Opportunistic Networks such as node energy, node buffer size, intermittent link or link capacity, and time-to-live of messages that must be addressed by the routing protocols. An efficient routing protocol should be straightforward, scalable, and capable of operating in low and high message load, and it should have a good delivery probability and low delay and overhead [9–10]. In this research work, we perform a comparative study of the performance of the most prominent routing protocols of the Opportunistic Networks, i.e. Direct Delivery [11], First Contact [12], Spray and Wait [13], Multischeme Spray and Wait [14], and Multischeme Adaptive Spray and Wait [15]. This section presents a brief description of these routing protocols.

18.2.1   DIRECT DELIVERY PROTOCOL

This is a basic and easy-to-follow routing protocol that forwards messages using a single-copy technique and does not require any network statistics information. A message is forwarded from a source to a destination only if the source is directly encountering the destination [11]. In this way, the delivery overhead is almost zero but delivery delay may be unlimited.

18.2.2   FIRST CONTACT PROTOCOL

This is a single-copy routing protocol, which means that each message has only one copy in the network. A node forwards a message to anyone of its neighbour randomly. If the node does not have any neighbour, it waits and forwards the message to its first contact node [12]. It performs very poorly because it chooses the next forwarder randomly without considering anything that leads the message to a wrong direction or dead end or loop.

18.2.3   SPRAY-AND-WAIT ROUTING PROTOCOL

The Spray-and-Wait [13] strategy was introduced to avoid blind flooding in the network with uncontrolled copies of the message and reduce resource usage in Epidemic routing [16]. It uses the flooding technique to forward a limited number of copies of a message. Compared to Epidemic routing protocol, Spray-and-Wait routing decreases the network overhead ratio and congestion. Routing is performed in two steps: spray and wait phase. In the spray phase, the source node forwards a fixed number of copies of a message – one copy to each encountered new node. In the wait phase, each node having a copy of the message can only deliver the message directly to the destination node only. The spray phase is identical to the Epidemic routing, while the wait phase is identical to the Direct Delivery routing. Another variation of the Spray-and-Wait routing is called Binary Spray-and-Wait routing in which the source nodes have N number of copies of the message, and when the first node comes in contact, the source forwards half of the copies (N/2) to it and keeps the remaining copies with itself. Similarly, each node having more than one copy of the message will hand over half of the copies to the encountering node if it does not have any copy of this message. When a node is left with just one copy of the message, then it can deliver that copy only to the destination node [13].

18.2.4   Multischeme Spray-and-Wait Protocol

To prevent a single spraying technique and blind forwarding by the source or relay nodes in Spray and Wait or Binary Spray and Wait, the Multischeme Spray-and-Wait routing was proposed in Ref. [14]. Multischeme Spray-and-Wait routing uses three spraying techniques during its spray phase; those are similar to the base protocol, similar to binary, and a new adaptive spray. However, the wait phase is the same as in the base protocol. Anyone of the available three spraying techniques will be used at a time on the basis of a threshold value decided on the basis of network conditions. In the adaptive spray phase, the number of copies to be forwarded to the encountering node is computed based on the delivery probability of the encountering node (like ProPHET [17]) to the destination of the message.

18.2.5   Multischeme Adaptive Spray-and-Wait Protocol

Multischeme Adaptive Spray-and-Wait routing works similarly to Spray-and-Wait routing and also considers node similarity and message delivery probability while forwarding copies of a message to the encountered nodes. The protocol forwards an asymmetric number of copies of a message to the encountered nodes on the basis of destination-aware utility and self-ware utility. The destination-aware utility indicates the capability of a node to forward the message towards a specific destination node. The self-aware utility indicates the capability of a node as a forwarder in general, irrespective of the message's destination. The delivery predictability is used as a destination-aware utility and node similarity is used as a self-aware utility. The basic idea is to give more message copies to the nodes having a higher possibility to deliver the message to the destination node [15].

18.3   ONE SIMULATOR

Opportunistic Network Environment (ONE) is a discrete event simulation engine with agents at its heart. A number of modules that implement the primary simulation functions are selected at each simulation step by the engine. Modelling of node movement, internode connections, routing, and message handling are among the core functions of the ONE simulator [18]. Visualisations, reporting, and post-processing tools are used to collect and analyse data. During the simulation time, reports generated by report modules are used to collect simulation outcomes. Report modules collect events from the simulation engine (such as message or connectivity events) and generate findings depending on them. The resulting data could be event logs that are subsequently processed by further external post-processing tools or aggregate data calculated in the simulator. The graphical user interface shows a visual representation of the simulation states including the nodes' positions, active connections, and message transmissions.

18.4 SIMULATION ENVIRONMENT SETUP

The ONE simulator [19] is used for the simulation of routing protocols and performance evaluation [20,21] of Direct Delivery, First Contact, Spray and Wait, Multischeme Spray and Wait, and Multischeme Adaptive Spray and Wait routing protocols of Opportunistic networks/DTN. Different types of mobility models are available in ONE simulator, namely Shortest Path Map–Based Movement, Map-Based Movement, Random Waypoint, and External movement models. Various parameters used to set up a common simulation environment for the performance evaluation of all routing protocols as well as protocol-specific parameters are shown in Tables 18.1 and 18.2.

TABLE 18.1
Common Simulation Environment Parameters

Parameters	Values
Simulation Area Size	4,500 m × 3,400 m
Simulation Time	3 hour
Message Size	500 kb – 1 Mb
Interface	Bluetooth
Interface Type	Simple Broadcast Interface
Bluetooth Transmit Range	10 m
Bluetooth Transmit Speed	250 kbps
Buffer Size	5 m, 10 m, 15 m, 20 m, 25 m, and 30 m
Number of Hosts	20 and 40
Message Generation Rate (message per min)	1
Message Time to Live	50, 100, 150, 200, 250, 300, 350, 400, 450, and 500
Mobility	MapRoute Movement
Movement Model	ShortestPathMapBasedMovement
Interval of Update	0.1
Interface	High Speed
High-Speed Interface Type	Simple Broadcast Interface
High-Speed Interface Transmit Speed	1,000
High-Speed Interface Transmit Range	10 m

TABLE 10.2
Protocol-Specific Simulation Parameters

Routing Algorithm	Parameter	Value
Direct Delivery	n/a	n/a
First Contact	n/a	n/a
Spray and Wait	Number of Copies (L)	6
Multischeme Spray-and-Wait Routing in Delay Tolerant Networks Exploiting Node Delivery Predictability	Number of Copies (L)	6
Multischeme Adaptive Routing Algorithm Based on Spray-and-Wait Delay-Tolerant Networks	Number of Copies (L)	6

18.5   PERFORMANCE METRICS

There are many performance matrices [7]; those may be analysed to evaluate the performance of routing protocols of the Opportunistic Networks. In this work, four main metrics are used for the performance evaluation, i.e. message delivery probability, latency, overhead, and composite metrics. Composite metrics show the overall effect of message delivery probability, latency, and overhead. A brief description of these performance metrics is as follows:

Message Delivery Probability: For a particular period of time, it is the ratio of messages received by the destination to messages sent by the source in the network.

$$\text{Message Delivery Probablility} = \frac{\text{No of Message at Destination Node}}{\text{No of Created Message by Source Node}} \quad (18.1)$$

Delivery Latency: For a particular period of time, it is the average time differences between the message delivery time and message creation time of all the successfully delivered messages.

$$\text{Average Delivery Latency} = \sum_{i \neq 1}^{n} \frac{(\text{Message Delivered Time}) - (\text{Message created Time})}{\text{No of Message delivered}}$$
$$(18.2)$$

Overhead Ratio: For a particular period of time, it is the ratio of the total number of messages relayed but not delivered and the number of messages successfully delivered.

$$\text{Overhead Ratio} = \frac{(\text{No of Message} - \text{No of Delivered Message})}{(\text{No of Delivered Message})} \quad (18.3)$$

Composite Metrics: A composite metric is used in addition to the above-stated usual metrics to get a comprehensive comparison of the protocols. The composite metric depicts how the primary metrics relate to each other. The composite metric gives credit for a higher delivery ratio, while penalising for both longer latency and higher overhead.

$$\text{Composite Metric (CM)} = \text{delivery ratio} \times \frac{1}{\text{latency}} \times \frac{1}{\text{overhead ratio}} \quad (18.4)$$

18.6 SIMULATION RESULTS

The performance analysis of the Direct Delivery, First Contact, Spray and Wait, Multischeme Spray and Wait, and Multischeme Adaptive Spray-and-Wait routing protocols of Opportunistic Networks/DTN has been performed. All routing protocols are simulated in a common simulation environment by using ONE simulator version 1.4.1. The performance is evaluated on the basis of metrics discussed in the previous section. Figures 18.1–18.8 show the results achieved for a simple "ShortestPathMapBasedMovement" mobility model. In all graphs, SnW abbreviation is used for Spray and Wait, MSnW is used for Multischeme Spray and Wait, and MASnW is used for Multischeme Adaptive Spray-and-Wait routing protocols.

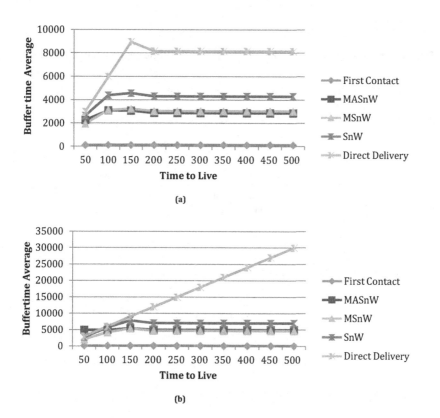

FIGURE 18.1 Buffer time average vs time to live. (a) Buffer size 5M, (b) Buffer size 10M.

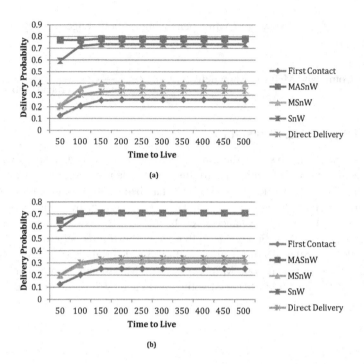

FIGURE 18.2   Delivery probability vs time to live. (a) Buffer size 5M, (b) Buffer size 10M.

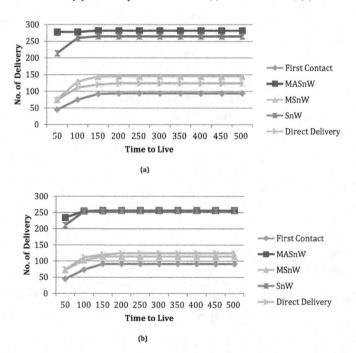

FIGURE 18.3   No. of deliveries vs time to live. (a) Buffer size 5M, (b) Buffer size 10M.

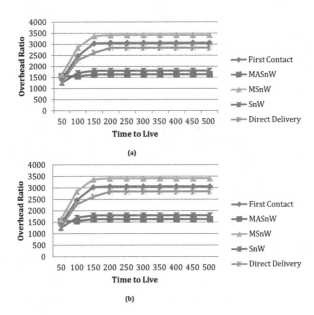

FIGURE 18.4   Overhead ratio vs time to live. (a) Buffer size 5M, (b) Buffer size 10M.

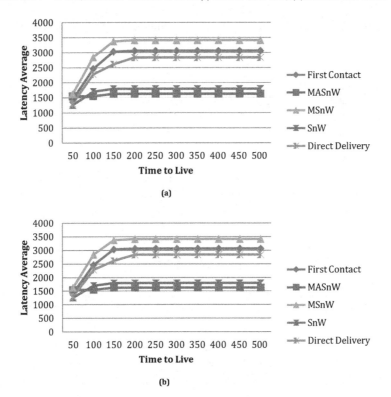

FIGURE 18.5   Latency average vs time to live. (a) Buffer size 5M, (b) Buffer size 10M.

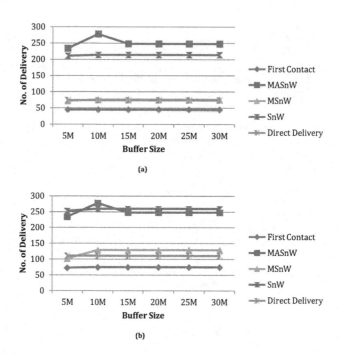

FIGURE 18.6 No. of deliveries vs buffer size. (a) Time to live 50, (b) Time to live 100.

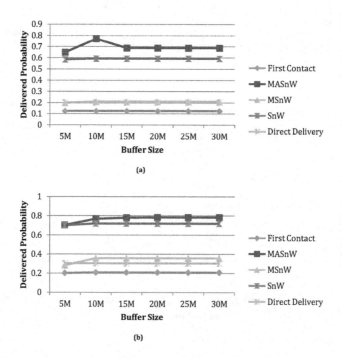

FIGURE 18.7 Delivered probability vs buffer size. (a) Time to live 50, (b) Time to live 100.

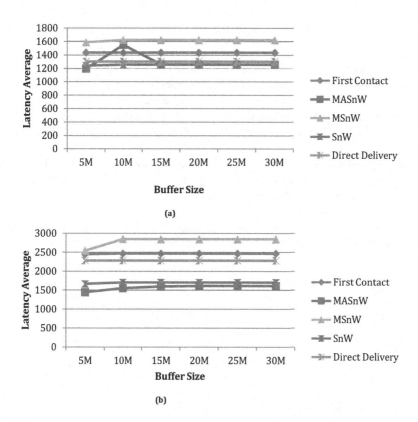

FIGURE 18.8   Latency average vs buffer size. (a) Time to live 50, (b) Time to live 100.

18.6.1   Impact of Different Time to Live

In Figures 18.1–18.5, it is clearly shown that the average buffer time and latency, delivery probability, and the number of delivery message increase with an increase in time-to-live of messages for all protocols at 3 hours of simulation time. The performance of all routing protocols is almost similar for the different times to live. Figures 18.1a and b show that the average buffer time is minimum in First Contact routing and is maximum in Direct Delivery in both cases of buffer size 5M and 10M. Other routing protocols' performance is in between Direct Delivery and First Contact routing protocols. Figures 18.2a,b and 18.3a,b show that respectively delivery probability and the number of deliveries are maximum in Multischeme Adaptive Spray and Wait and are minimum in First Contact in both cases of buffer size 5M and 10M. Figures 18.4a and b show that the overhead ratio is the highest in Multischeme Adaptive Spray and Wait and lowest in the Direct Delivery routing protocol. The reason for the low overhead ratio of Direct Delivery is discussed in Section 18.2.1. Figures 18.5a and b show that the average latency is the lowest in Multischeme Adaptive Spray and Wait and highest in Multischeme Spray-and-Wait routing protocols in both cases of buffer size 5M and 10M.

18.6.2 Impact of Different Buffer Sizes

In Figures 18.6–18.10, it is clearly shown that the average buffer time and latency, delivery probability, and the number of delivered messages increase with an increase in the message time to live for all protocols at 3 hours of simulation time. The performance of all routing protocols is almost similar to different buffer sizes. Figures 18.6a,b and 18.7a,b show that respectively the number of delivery and delivery probability are the highest in Multischeme Adaptive Spray and Wait and lowest in First Contact in both cases of time to live 50 and 100. Figures 18.8a and b show that average latency is the lowest in Multischeme Adaptive Spray and Wait and highest in Multischeme Spray-and-Wait routing in both cases of time to live 50 and 100. Figures 18.9a and b show that the overhead ratio is the highest in First Contact and lowest in Direct Delivery routing protocol. Figures 18.10a and b show that average buffer time is the lowest in First Contact and highest in Direct Delivery in both cases of time to live 50 and 100.

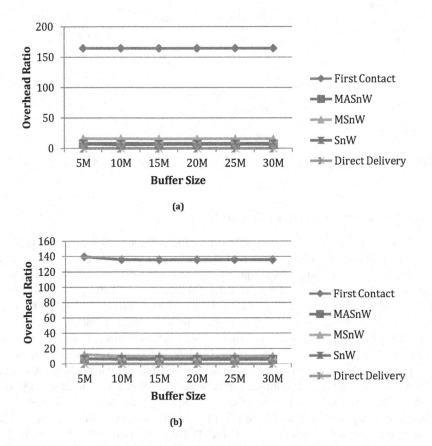

FIGURE 18.9 Overhead ratio vs buffer size. (a) Time to live 50, (b) Time to live 100.

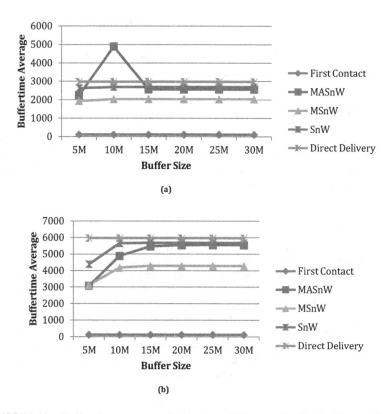

FIGURE 18.10   Buffer time average vs buffer size. (a) Time to live 50, (b) Time to live 100.

18.7 CONCLUSION

Routing is one of the main issues in Opportunistic Networks [22–26]. For the selection of a routing protocol for a network or an application, it is very important to know the performance of the existing routing protocols in diverse simulation environments. In this work, we have simulated and investigated the performance of routing protocols designed for Opportunistic Networks/DTNs, i.e. Direct Delivery, First Contact, Spray and Wait, Multischeme Adaptive Spray and Wait, and Multischeme Spray-and-Wait routing protocols. One mobility model "shortest path map-based model" was used for simulations. It is clearly shown by the simulation results that the average buffer time and latency, delivery probability, and the number of delivery message increase with an increase in time-to-live of messages for all protocols at 3 hours of simulation time. Other interpretations of simulation results are self-explanatory. In our future work, we will develop new context-aware routing protocols for Opportunistic Networks and evaluate their performance against the existing routing protocols. Opportunistic networking concept can also be used to provide distributed computing in intermittently connected Ad hoc Networks.

Therefore, developing lightweight failure detection algorithms [27–29] to monitor the failures in such networks may be an emerging research direction. Exploring issues and creating solutions to resolve them for Opportunistic Networks made up of Internet-of-Things devices [30–32] are also in trend.

ACKNOWLEDGEMENT

This work is part of the research work funded by "Seed Grant to Faculty Members under IoE Scheme (under Dev. Scheme No. 6031)" awarded to Anshul Verma and "DST-Science and Engineering Research Board (SERB), Government of India (File no. PDF/2020/001646)" awarded to Pradeepika Verma.

REFERENCES

1. A. Verma, K. K. Pattanaik, and A. Ingavale, 'Context-based routing protocols for oppnets', in *Routing in Opportunistic Networks*, edited by I. Woungang, S. K. Dhurandher, A. Anpalagan, and A. V. Vasilakos, Springer, New York, 2013, pp. 69–97.
2. I. Woungang, S. K. Dhurandher, A. Anpalagan, and A. V. Vasilakos (Eds.), *Routing in Opportunistic Networks*, Springer, New York, 2013.
3. P. Mitra and C. Poellabauer, 'Opportunistic routing in mobile ad hoc networks', in *Routing in Opportunistic Networks*, edited by I. Woungang, S. K. Dhurandher, A. Anpalagan, and A. V. Vasilakos, Springer, New York, 2013, pp. 145–178.
4. F. Warthman, 'Delay- and Disruption-Tolerant Networks (DTNs) - A turotial', *Ipnsig*, pp. 1–35, 2015, [Online]. Available: http://ipnsig.org/wp-content/uploads/2015/09/DTN_Tutorial_v3.2.pdf.
5. Y. Li, P. Hui, D. Jin, and S. Chen, 'Delay-tolerant network protocol testing and evaluation', *IEEE Commun. Mag.*, vol. 53, no. January, pp. 258–266, 2015.
6. S. K. Kushwaha, A. Kumar, and N. Kumar, 'Routing protocols and challenges faced in ad hoc wireless networks', *Adv. Electron. Electric Eng*, vol. 4, no. 2, pp. 207–212, 2014.
7. C. Prabha and R. Kumar, 'Performance comparison of routing protocols in opportunistic networks', in *Proceedings of International Conference on Research in Management& Technovation 2020*, 2020, vol. 24, pp. 87–90, doi: 10.15439/2020km16.
8. S. K. Dhurandher, D. K. Sharma, I. Woungang, S. Gupta, and S. Goyal, 'Routing protocols in infrastructure-based opportunistic networks', in *Routing in Opportunistic Networks*, edited by I. Woungang, S. K. Dhurandher, A. Anpalagan, and A. V. Vasilakos. Springer, New York, 2013, pp. 125–144.
9. T. Prodhan, R. Das, H. Kabir, and G. C. Shoja, 'TTL based routing in opportunistic networks', *J. Netw. Comput. Appl.*, vol. 34, no. 5, pp. 1660–1670, 2011, doi: 10.1016/j.jnca.2011.05.005.
10. S. J. Anandh and E. Baburaj, 'Energy efficient routing technique for wireless sensor networks using ant-colony optimization', *Wirel. Pers. Commun.*, vol. 114, no. 4, pp. 3419–3433, 2020, doi: 10.1007/s11277-020-07539-0.
11. M. Alajeely, A. Ahmad, and R. Doss, 'Comparative study of routing protocols for opportunistic networks', in *Proceedings of International Conference on Sensor Technology (ICST)*, 2013, pp. 209–214, doi: 10.1109/ICSensT.2013.6727644.
12. S. Jain, K. Fall, and R. Patra, 'Routing in a delay tolerant network', in *Proceedings of the 2004 Conference on Applications, Technologies, Architectures, and Protocols for Computer Communications*, August 2004, Portland, OR, pp. 145–158.

13. T. Spyropoulos, K. Psounis, and C. S. Raghavendra, 'Spray and wait: An efficient routing scheme for intermittently connected mobile networks', in *Proceedings of the ACM SIGCOMM 2005 Workshop on Delay-Tolerant Networking, WDTN 2005*, 2005, pp. 252–259, doi: 10.1145/1080139.1080143.
14. S. M. A. Iqbal, 'Multischeme spray and wait routing in delay tolerant networks exploiting nodes delivery predictability', in *2012 15th International Conference on Computer and Information Technology (ICCIT)*, 2012, pp. 255–260, doi: 10.1109/ICCITechn.2012.6509722.
15. J. Li, S. Jiang, Y. Song, J. Xu, and Y. Wang, 'A multi-scheme adaptive routing algorithm based on spray and wait for delay tolerant networks', *Int. J. Smart Sens. Intell. Syst.*, vol. 8, no. 4, pp. 2136–2158, 2015, doi: 10.21307/ijssis-2017-846.
16. M. J. F. Alenazi, Y. Cheng, D. Zhang, and J. P. G. Sterbenz, 'Epidemic routing protocol implementation in NS-3', in *ACM International Conference Proceeding Series*, no. May, 2015, pp. 83–90, doi: 10.1145/2756509.2756523.
17. A. Lindgren, A. Doria, and O. Schelén, 'Probabilistic routing in intermittently connected networks', *Lect. Notes Comput. Sci. (including Subser. Lect. Notes Artif. Intell. Lect. Notes Bioinformatics)*, Fortaleza, Brazil, vol. 3126, pp. 239–254, 2004, doi: 10.1007/978-3-540-27767-5_24.
18. A. Keränen, J. Ott, and T. Kärkkäinen, 'The ONE simulator for DTN protocol evaluation', *Proceedings of the 2nd International Conference on Simulation Tools and Techniques*, Rome, March 2–6, 2009, pp. 1–10.
19. A. Keränen, T. Kärkkäinen, and J. Ott, 'Simulating mobility and DTNs with the ONE', *J. Commun.*, vol. 5, no. 2, pp. 92–105, 2010, doi: 10.4304/jcm.5.2.92-105.
20. R. Saxena, V. Rishiwal, and O. Singh, 'Performance evaluation of routing protocols in wireless sensor networks', in *Proceedings of 2018 3rd International Conference on Internet Things Smart Innovation and Usages, IoT-SIU 2018*, no. July, 2018, doi: 10.1109/IoT-SIU.2018.8519933.
21. S. K. Dhurandher, D. K. Sharma, I. Woungang, and H. C. Chao, 'Performance evaluation of various routing protocols in opportunistic networks', in *2011 IEEE GLOBECOM Workshops (GC Workshops)*, 2011, pp. 1067–1071, doi: 10.1109/GLOCOMW.2011.6162342.
22. A. Verma, M. Singh, K. K. Pattanaik, and B. K. Singh, 'Future networks inspired by opportunistic networks', in *Opportunistic Networks*, edited by Khaleel Ahmad, Nur Izura Udzir, Ganesh Chandra Deka, Chapman and Hall/CRC, Boca Raton, FL, 2018, pp. 230–246.
23. A. Verma and K. K. Pattanaik, 'Routing protocols in opportunistic networks', in *Opportunistic Networking*, CRC Press, Boca Raton, FL, 2017, pp. 123–166.
24. M. Singh, P. Verma, and A. Verma, 'Security in opportunistic networks', in *Opportunistic Networks*, CRC Press, Boca Raton, FL, 2021, pp. 299–312.
25. A. Verma and D. Srivastava, Integrated routing protocol for opportunistic networks. arXiv preprint arXiv:1204.1658, 2012.
26. A. Verma, P. Verma, S. K. Dhurandher, and I. Woungang, (Eds.), *Opportunistic Networks: Fundamentals, Applications and Emerging Trends*, CRC Press, Boca Raton, FL, 2021.
27. A. Verma and K. K. Pattanaik, 'Failure detector of perfect P class for synchronous hierarchical distributed systems', *Int. J. Distrib. Syst. Technol. (IJDST)*, vol. 7, no. 2, pp. 57–74, 2016.
28. A. Verma, M. Singh, and K. K. Pattanaik, 'Failure detectors of strong S and perfect P classes for time synchronous hierarchical distributed systems', in *Applying Integration Techniques and Methods in Distributed Systems and Technologies*, edited by Kecskemeti, Gabor, IGI Global, 2019, pp. 246–280.

29. B. Chaurasia and A. Verma, 'A comprehensive study on failure detectors of distributed systems', *J. Sci. Res.*, vol. 64, no. 2, 2020, pp. 250–260.

30. A. Verma, P. Verma, Y. Farhaoui, and Z. Lv (Eds.), *Emerging Real-World Applications of Internet of Things*, CRC Press (Taylor and Francis), Boca Raton, FL, 2022.

31. B. Rudra, A. Verma, S. Verma, and B. Shrestha (Eds.), *Futuristic Research Trends and Applications of Internet of Things*, CRC Press (Taylor and Francis), Boca Raton, FL, 2022.

32. S. Srivastawa, A. Verma, and P. Verma, 'Fundamentals of internet of things', *Futuristic Research Trends and Applications of Internet of Things*, edited by B. Rudra. A. Verma, S.Verma and B. Shrestha. CRC Press, Boca Raton, FL, 2022, pp. 1–30 .

19 Reduce the Privacy and Security Concerns of Current Social Media Platforms Using Blockchain Technology

Muneshwara M. S., Swetha M. S.,
Anand R., and Anil G. N.
BMS Institute of Technology and Management

CONTENTS

19.1 INTRODUCTION

Social media has become a key influence in the lives of many people in the society today. It has provided a way for people to keep in touch with other people in the world and also a means to share informative and interesting content to our loved ones. It has also become a platform used by companies to engage with their customers. But one of the problems that have risen over time is the huge monopolization of data as seen by huge corporates that make millions off of unsuspecting user data.

DOI: 10.1201/9781003320333-19

The study models a peer-to-peer social media network that makes use of blockchain in anonymizing user data interactions. A blockchain is defined as a decentralized digital ledger of transactions that is distributed, most commonly public, and can be kept track across multiple node computers. A blockchain is essentially a digital ledger that is distributed, publicly available, decentralized, and used to keep track of transactions that take place through various computers. Any altercation of these transaction records cannot take place without changing out all subsequent blocks; therefore, there is no central point of vulnerability that can be exploited in the network. A security method that is widely used in blockchain is the public-key cryptography. The social media network is spread across thousands of devices working together instead of the traditional centralized database approach. Information privacy is the act of maintaining the relationship between the collection of data and the distribution/spreading of data. This may include conforming to the public expectation of privacy while also dealing with any legal issues related to dissemination of public data. Each post in the media can be tracked using the blockchain where the user can track where the data they posted has traveled. Monetization can be achieved through user's posts on the social media. Here, the user makes money in cryptocurrency for each view on his/her post. Hence, in the end, a decentralized social media platform is obtained that allows us to secure data and maintain privacy of the user as well.

The spreading of rumors that take place at a large scale can also lead to serious social and economic damages. As it is famously said, false information spreads faster than any virus in the world. Each post made public on the social media will be analyzed by a machine learning algorithm running on each node to determine which parts of it are fake and which are real. Spreaders and posters of false information are penalized with a reduction in their credit score on the network making their posts less visible on the network.

19.2 LITERATURE REVIEW

Murimi et al. investigate the various methods that have been developed by social media sites to allow users to create and share content. Various social network sites exist today, and each of these sites offers different ways for users to express themselves anonymously or not and be part of a community of users around the world. This rise has led to the challenge of how content is disseminated and accessed fairly with the user's acknowledgment. Here, the paper talks about a framework that the author has proposed to provide a method to generate value for user-created content by making use of blockchain. It discusses how the framework can be used in various fields and challenges that may come overtime when implementing such a blockchain-powered social media network [1]. Li and Palanisamy look into Steemit, which is a reward-based social media platform that allows users to share content and rank them based on votes of other users in the network. It provides cryptocurrency as a reward to users that contribute to popular posts in the network. Though on the outside it may seem that this reward system is used to promote users to produce higher quality, network transfer suggests that majority of the supply is taken away by bots, which seem to suggest that there is a huge misuse and manipulation of Steemit platform's reward system. One core limitation this paper has is it does not take into account data lost due to Web scraping [2].

Kumari and Singh look into the major worries that do relate with respect to social media that include the security and the privacy of users. The important challenge is to protect user's personal data in accordance with the laws and policies that have been instated with respect to data rights. There should not be any disclosure of any private data with respect to a user in the social media network. The core challenge with respect to privacy is usually seen in one direction because the privacy of a person is exploited not only on the outside but can also be caused due to negligence of the users themselves. These users are not aware of how data provided by them is made use of. Such methods and awareness are to be made for protecting user's personal data. There is also growing trend where the social networking website themselves sells the user data to advertisement companies to make up the money that they invested. Many users don't realize this process, and their data ends up with the vast range of advertising companies. These companies can advertise selectively and can also build a monopoly for their products. By providing a reward system, people have a sense of satisfaction as they are paid for the data they provide. They can also view which company acquires data and how they use it. This provides more transparency compared to current system [3]. Talwar et al. provide an insight on how sharing takes place as well as the motives for people to share. [1] It also discusses these behaviors with association with a framework and third-person effect hypothesis [4]. Waseem discuss how social media has evolved and how it has affected us in different ways in our lives in positive and negative ways. It lists out all social media networks that are present today and gives brief descriptions of them. It also discusses the effects it has in different ways in our lives. A limitation would be it being a rather less technical paper than usual and majority situations discussed in the paper are highly hypothetical with lack of study-based evidence [5].

Zhang et al. discuss and study fake news spread on various parameters and highly detailed as well. It looks into algorithms and methodologies that can be used for detecting fake news. It also introduces a fake news credibility inference model and compares its results with other existing models [6]. Aldwairi and Alwahedi discuss fake news as well as bring up the concept of clickbait and how it has become prevalent in the current Internet age. It proposes a solution to help dissipate this problem, but the solution is too simple and hugely constrained to that dataset provided in the paper [7]. Steni and Sreeja present a survey of fake news detection on the basis of community opinion to various users as well as the post itself with the help of machine learning algorithms [8].

Chen et al. look to discuss how blockchain could be used to tame and control the spread of false information or rumors. It also discusses the possible new age of social media networks where blockchain can become a key technology in producing networks with trust and also provide a secure way of user data dissemination [9]. Freni et al. talk about the issues with social media today in relation to user data security and talk about how blockchain can be used to resolve such issues. It also discusses a token-based system to provide incentives to provide a reward of sorts to users to gain an incentive for the data they share [10].

Casino looks to provide a review of how blockchain can be applied in developing solutions to problems that persist in various domains. It investigates how blockchain technology can be used to drastically improve today's practices and how it is used today. It also provides a pathway of how blockchain can be a good fit in each of the applications [11]. Zhang et al. survey and discuss few cryptocurrency incentive schemes that exist in blockchain-based solutions and have also compared these schemes. It also mainly discusses these schemes around Steemit that is a major blockchain-based social network [12]. Blockchain has been a technology being talked about a lot these days. Pfeiffer et al. provide a brief look into how blockchain technologies could be used to influence as well as create social networks and give an overview of currently present platforms, which facilitate blockchain technologies and inculcate social media with it [13]. Ciriello et al. look into how blockchain has allowed us to enable and control regular social media practices. It shows how Steem offers various options of social media enabling as well as social media constraining option and how it can have contradictory effect on regular social networking practices [14]. Xie investigates how incentives in blockchain social networks and how the hierarchy of the users for sharing in the network effect the quality of information that is shared on the social network. It also discusses how higher visibility of a post makes it spill over to other cryptocurrency markets easier [15].

19.3   EXISTING SYSTEM

Various social network sites exist today, and each of these sites offers different ways for users to express themselves anonymously or not and be part of a community of users around the world [16] (Figure 19.1).

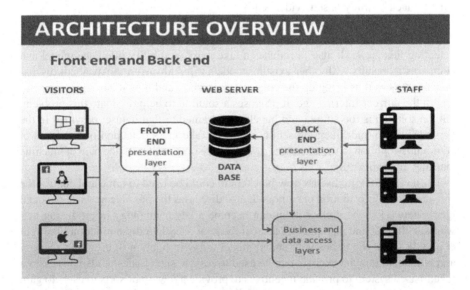

FIGURE 19.1   Facebook architecture overview.

As seen above there is a traditional three-layer centralized architecture being used by many of the current social media networks. This is a typical client/server architecture where the server exposes certain endpoints through which the user can access the resources provided by the server. Here, the server is in control of the resources received by the client. Since the servers are owned by corporations whose main goal is to make profit, there is a high chance of user data being used for targeted advertising, selling bulk data for profits, etc.

Limitations

- The media network is run by servers owned by large corporations and not by the user themselves.
- There are hundreds of cyber-attacks each day on the centralized servers, which can leak huge amounts of user data and hence cause harm.
- User data is misused by large corporations where the data is used to predict user behavior and modify it for making profits. The biggest scandal that had come to the public's attention was the Facebook–Cambridge Analytica data scandal. This was a scandal that emerged where personal data of many Facebook users was found to be collected without any permission by the Cambridge Analytica and was mainly used for political promotion and advertising. This is known to be one of the largest leaks of Facebook history.
- A lot of Web scraping happens where user data is stolen and used illegally to train machine learning models.
- The user who drives the social media does not make any profit, whereas the companies hosting the media make billions of dollars in revenue selling and using data.
- There is no sufficient mechanism to find and reduce false information such as false news and rumors in the network.

19.4 SYSTEM DESIGN

19.4.1 Blockchain Design

A blockchain is a distributed and decentralized digital ledger, which is most often made public. It is used to keep track transactions that happen across many computers. Since the blockchain is expanded in such a manner, any record that is considered part of the network cannot be altered, without making those alterations to all subsequent blocks (Figure 19.2).

Blockchain security methods make use of public-key cryptic techniques [17]. A public key is a string of random letters and numbers and acts as an address for the nodes on the blockchain. Value tokens sent across the network are recorded as belonging to that address. A private key acts as a counterpart to a corresponding public key and is used by the owner to access his/her confidential data. RSA is extensively used in the blockchain to create a wallet for each user.

RSA also known as Rivest–Shamir–Adleman algorithm is a public-key encryption/decryption algorithm and is widely used in networking to transfer data from one

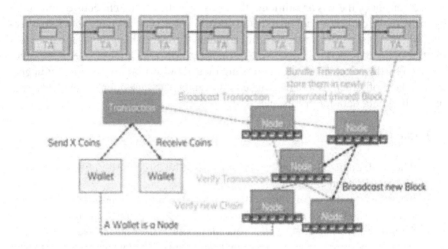

FIGURE 19.2 Blockchain architecture.

place to another securely. The public key is also known as the encryption key and its available for all users to view and is different from the private key, also known as the decryption key, which is kept as a secret. Messages that are encrypted [18] using the public key can only be decrypted with the help of a private key. It is almost impossible to find a private key given a public key and this is at the heart of the algorithm's security. RSA has one disadvantage that it is slow and is hence used to transfer symmetric keys, which can then be used for transfer later. Here, RSA can be used to generate public–private key pairs that are uniquely used to identify users in the network. Each public–private key pair is also known as a wallet, and hence, a user can be identified in terms of his wallet. The blockchain stores transactions that have occurred between two wallets. Whenever a user views an anonymized and monetized post in his feed, there will be a transaction from the latter to the person who posted the content. This is the incentive system used in our model.

19.4.2 Incentives/Gains for the User

The rate awarded to the user on receiving hits on his/her post will be decided by an algorithm that takes the following factors into account such as exposure of the post, amount of false information in it, and credit score of the user [19] (determined by the network based on his past activity). It can be as simple as a linear combination of a variety of factors.

$$\text{Incentive earned} = w_0 + (w_1 * \text{no\_of\_views}) + (w_2 * \text{credit\_score})$$

w_0 is the base incentive earned on creating a post on the social media and can range from 0 to 100. w_1 and w_2 are weighted values that determine the contribution of each of the factors into the incentive earned. For example, business accounts may have a higher value for w_1 than w_2.

19.4.3 CREDIT SCORES

Credit score is a real-valued number. Credit scores are a reflection of the user behavior on the network. Users who share malicious content or engage in practices that violate the policies of the social media network will be penalized leading to a reduction in their credit scores.

$$\text{Reduction in credit score} = \left(\text{no of reports for malicious content} * a\right)$$

a is a variable and can vary in order to reduce or increase the reduction of credit score.

For example, user X has 100 credits and engages in the malicious practices, which are then reported by 3 users. Then, the reduction in credit score would be 3a. Let us say that the post was from an account that has a history of penalties; in this, case the value of a would be high (say 10), and hence, the user's credit score would reduce to 70.

A user can make use of the false information detection mechanism, which can aid him in deciding whether he must report a post or not.

19.5 ARCHITECTURAL DESIGN

The proposed architecture can be viewed as a stack of 3 layers each equally important for the proper functioning of the entire system:

The Blockchain Layer: Where the transactions of the user data are stored in a systematic matter where data manipulation can be easily detected.

The Peer Nodes Layer: Where each peer node has a copy of the blockchain, these nodes run the machine learning algorithms and communicate with each other to share block data.

The User Layer: This layer is also known as the wallet in blockchain jargon, this layer provides the user interface, which communicates with the peer nodes. The user can either use the services of another peer node or be a peer node himself (Figure 19.3).

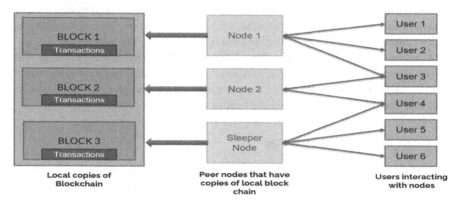

FIGURE 19.3 Proposed system architecture.

The user can create a post and choose to anonymize and track the data (basically sell it for a profit in crypto). The blockchain in itself has its own currency, which is awarded to the user whenever there are views on his post. If a user chooses to anonymize the post, the node will then remove his name and add his/her public key to the post. Whenever the post is viewed by someone, there will be a transaction on the blockchain from the viewer's public key to the public key of the original poster, which cannot be traced back to the latter (the original posters name will not be known only his public key). The original poster will be the only one who can track the transactions on the blockchain using his/her private key and hence know exactly who has his/her data. Hence, the user can post his information on the media and track where it travels while making money out of it. If the user feels his data is being used illegally, he can immediately report the transaction and the corresponding holder's account will be put under inspection on the peer-to-peer network, which can reduce his credit score.

The currency gained by the user from the post can be used by the user to view further posts or buy data from other posts on the social media network, which are also recorded as transactions on the blockchain.

The rate awarded to the user will be decided by an algorithm that takes the following factors into account such as exposure of the post, amount of false information in it, and credit score of the user (determined by the network based on his past posts).

Detection of false information in the evaluation of the rate, an analysis of some of the major reasons for the spread of false information which are false news is more inclined toward a user's beliefs, peer approval and political parties pushing their agenda. Posts made public on the social media will be analyzed by a natural language processing algorithm running on each node to determine parts of it that are real [20] and parts that are fake. The viewer of a post can click fact checker button, and parts fake will be highlighted in red, while parts that seem real will be highlighted in green. The original poster will then be penalized on his credit score, and any subsequent viewers who want to share it will be given a warning to not share such a post. If they continue to do so, they too will also receive a penalty on their credit score.

Users with lower credit scores will have their post visibility limited and eventually phased out of the network. The network will never run out of space or compute as nodes communicate with each other and find more nodes to share load with.

A machine learning regression algorithm predicts the requirement in storage that may arise each day, and hence, the system is always prepared to face expansion on its own. There will be nodes called sleeper nodes on the system [21], which will be activated and deactivated based on the needs of the network.

19.6   COMPONENT DESIGN

Blockchain nodes are the core component of the social media network [22]. Each node that is connected in the blockchain consists of 2 core components as shown in Figure 19.4.

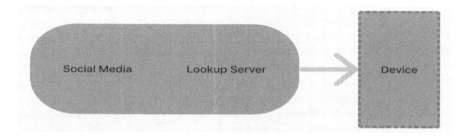

FIGURE 19.4 Design of each node of the network.

1. **Social Media**

    The social media server is responsible for the following:
    - Deliver the user interface.
        - Handle requests from the user interface. These requests can be of various types, i.e., login/signup, data retrieval, and uploads.
        - The data is stored in a NoSQL database under the collection social media.
        - The distributed database appears as a single central database to the social media server that it can query.
        - The social media server has no idea about other peers and functions like it is the only server that is active in the environment.

2. **Lookup Server**

    The lookup server lies at the heart of each peer node and is responsible for establishing communication between the nodes; it also performs the following functions:
    - **Register Itself with the Network:** The lookup server registers itself with the master lookup server via a register request. The master lookup returns the set of peers from its database. The lookup server then stores these peers in its database and then registers itself with each of the peers received via a register request.
    - **Send a Heartbeat When It Comes Online after a Failure:** The node may have gone down for many reasons and hence would have lost connectivity with the network. Any peer connected to the node going offline will mark this node as an inactive node in its database. It is the responsibility of the lookup server to send a heartbeat to all the peers in its database when it comes back online.
    - **Retrieve Data Not Present Locally:** Whenever the social media receives a request for data not present in its local database, the lookup server takes the responsibility to locate the data. It requests each peer node in its database for the data; if the node has the data, it returns an acknowledgment.

    The lookup server then retrieves this data from the address of that node. Once the data is retrieved from a node in the network, [23] the

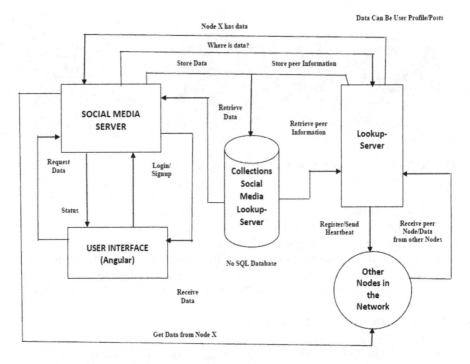

FIGURE 19.5 Dataflow diagram of each peer in the network.

lookup server places this data in the social media collection of the database. This data is then retrieved from the collection by the social media
server and given to the user interface for display (Figure 19.5).

19.7 RESULT AND DISCUSSION

The final application allows each user to create a dedicated account using an email
and password. Once the user is logged in, the user is greeted with his post list, which
includes posts or rest of the peers on the network where the user can interact with
rest of the posts with likes and comments. Any malicious or spam-like posts can be
reported reducing the credit score of that user.

Each post can be either public or private based on the options given while creating
a post. All private posts need to be "paid" for if a user wants to use or view the post
with cryptocurrency of the network. Users can mine cryptocurrency as well.

The resultant screenshots of the system are as shown below (Figures 19.6–19.11):

- Login Page
- Register Page
- My Profile Page
- Post List and Create New Post
- Report Post
- Credit Score Comparison

FIGURE 19.6   Login page with create account button.

FIGURE 19.7   Register page with account creation form.

FIGURE 19.8   My profile page.

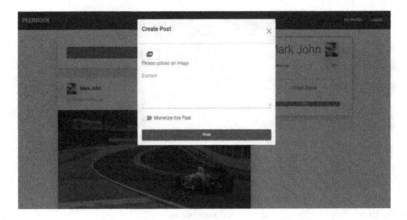

FIGURE 19.9 Modal to create new post.

FIGURE 19.10 Modal to repost post.

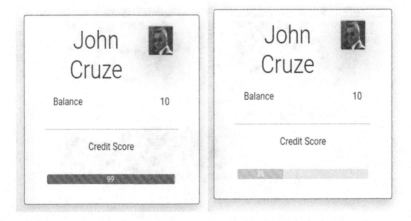

FIGURE 19.11 Comparison between good credit score and poor credit score.

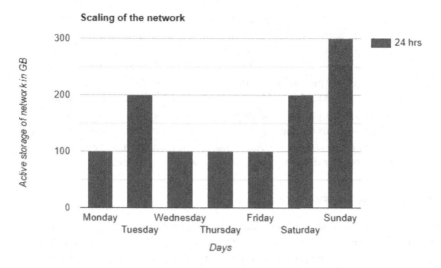

FIGURE 19.12 Graph showing scalability in the storage space.

The big picture here is that this is a study on a peer-to-peer social network that is decentralized with the help of blockchain. Since blockchain is a distributed digital ledger [24] and is most often available to the public, any record in the network (in our case transactions of data) cannot be modified or altered without making those modifications to subsequent blocks. This significantly reduces the risk of cyber-attacks, which results in a more secure system.

19.7.1 SCALABILITY

The graph shown in Figure 19.12 shows a trend in the storage of the network over a period of 1 week. Tuesday being a public holiday sees a large increase in the activation of sleeper nodes for more storage as there are a lot of posts coming into the network. This then reduces to 50 GB in the next 3 days. Note: This does not mean that the remaining 150 GB of data was lost; it simply means that the nodes containing this data were never activated nor did they get any requests for this data as the user or any of their viewers were never online. Again on the weekends where there is an increase in the traffic, the sleeper nodes are activated in order to store/retrieve the data.

19.8 CONCLUSION AND FUTURE ENHANCEMENT

The system "**Peer-To-Peer Social Media Network Using Blockchain**" is developed and tested successfully and satisfies all the requirements of the user. The goals that have been achieved by the developed system are as follows:

- Distributed social network that scales itself based on need and usage, working social media network that allows users to share posts, provide security and privacy for users using the blockchain, reward users of the network that choose to share their data, and minimize spam and false information through the network. With the help of a blockchain, a network that is self-scaling and distributed was implemented. The system [25] also provides a model to limit the propagation of false information. The users on the network are rewarded for sharing their data through the built-in currency. Data privacy was achieved by replacing user identities with public keys for the transactions on the blockchain. The outcome of building a distributed social network that scales itself based on need and usage, a network owned by the user and not by any corporation, and data privacy and [26] security by isolating user data from user identity was also achieved.

19.8.1 SCOPE FOR FUTURE ENHANCEMENT

Study the reports received from various users in the network and make use of machine learning algorithms to learn what kind of posts are spreading false information and prevent such posts from spreading in the network. Implement a way for people to connect with one another by allowing them to add friends to their network and separate the regular public feed from the private social network feed.

REFERENCES

1. R. M. Murimi, "A blockchain enhanced framework for social networking", *Ledger*, vol. 4, 2019.
2. C. Li, B. Palanisamy, "Incentivized blockchain-based social media platforms: A case study of steemit", *Proceedings of the 10th ACM Conference on Web Science*, 2019, pp. 145–154. doi: 10.1145/3292522.3326041.
3. S. Kumari, S. Singh, "A critical analysis of privacy and security on social media", 2015 *Fifth International Conference on Communication Systems and Network Technologies*, Gwalior, MP, India, April 2015, pp. 602–608, IEEE.
4. S. Talwar, A. Dhir, D. Singh, G. Virk, J. Salo, (2020). Sharing of fake news on social media: Application of the honeycomb framework and the third-person effect hypothesis. *Journal of Retailing and Consumer Services*, 57, 102197. doi: 10.1016/j.jretconser.2020.102197.
5. W. Akram, (2018). A study on positive and negative effects of social media on society. *International Journal of Computer Sciences and Engineering*, 5. doi: 10.26438/ijcse/v5i10.351354.
6. J. Zhang, B. Dong, P. S. Yu, "FakeDetector: Effective fake news detection with deep diffusive neural network", *2020 IEEE 36th International Conference on Data Engineering (ICDE)*, 2020, pp. 1826–1829, doi: 10.1109/ICDE48307.2020.00180.
7. M. Aldwairi, A. Alwahedi, (2018). Detecting fake news in social media networks. *Procedia Computer Science*, 141, 215–222. doi: 10.1016/j.procs.2018.10.171.
8. T. S, Steni, P. S, Sreeja, (2020). Fake news detection on social media-a review. *Test Engineering and Management*, 83, 12997–13003.
9. Y. Chen, Q. Li, H. Wang, "Towards trusted social networks with block chain technology", *Symposium on Foundations and Applications of Blockchain Proceedings*, University of Southern California, Los Angeles, California, 2018.
10. P. Freni, E. Ferro, G. Ceci, "Fixing social media with the blockchain", *Proceedings of the 6th EAI International Conference on Smart Objects and Technologies for Social Good*, 2020. pp. 175–180. doi: 10.1145/3411170.3411246.

11. F. Casino, T. K. Dasaklis, C. Patsakis, (2019). A systematic literature review of block-chain-based applications: Current status, classification and open issues. *Telematics and Informatics*, 36, 55–81.

12. R. Zhang, J. Park, R. Ciriello, (2019). "The differential effects of cryptocurrency incentives in blockchain social networks", *Pre-ICIS Workshop on Blockchain and Smart Contract (SIGBPS2019)*, Munich, Germany.

13. A. Pfeiffer, S. Kriglstein, T. Wernbacher, S. Bezzina, "Blockchain technologies and social media: A snapshot", *ECSM 2020 8th European Conference on Social Media*, 2020. doi: 10.34190/ESM.20.073.

14. R. Ciriello, R. Beck, J. Thatcher, (2018). The Paradoxical Effects of Blockchain Technology on Social Networking Practices. Available at SSRN 3920002.

15. P. Xie, (2020). The effect of incentive hierarchy system of social media in the delivery of quality information. *Journal of International Technology and Information Management*, 28(4), pp. 1–26.

16. M. S. Muneshwara, H. Vallae, M. S. Swetha, M. Thungamani, "Comparison on hyper ledger fabric and hyper ledger composer of block chain technology", *2019 International Conference on Intelligent Computing and Control Systems (ICCS)*, Madurai, India, 2019.

17. M. S. Swetha, M. Thungamani, (2019). A novel approach to secure mysterious location based routing for manet. *International Journal of Innovative Technology and Exploring Engineering (IJITEE)*, 8(7), pp. 2587–2591.

18. G. Han, L. Liu, N. Bao, J. Jiang, W. Zhang, J. Rodrigues, (2017). AREP: An asymmetric link-based reverse routing protocol for underwater acoustic sensor networks. *Journal of Network and Computer Applications*, 92, 51–58.

19. G. Sarraf, M. S. Swetha, (2020). Intrusion prediction and detection with deep sequence modeling. *Communications in Computer and Information Science*, vol 1208. Springer, Singapore, ISBN: 978-981-15-4825-3.

20. M. S. Muneshwara, A Lokesh, M. S. Swetha, MThunagmani, "Ultrasonic and image mapped path finder for the blind people in the real time system", *2017 IEEE International Conference on Power, Control, Signals and Instrumentation Engineering (ICPCSI)*, Chennai, India.

21. M. S. Muneshwara, B. R. Rajendra, Intelligent Robot Positioning System (IRPS) for tracing the contemporary location. *IAETSD Journal for Advanced Research in Applied Sciences Scientific Journal Impact Factor -5.2 Indexed by: Thomson Reuters' Research ID: H-2404-2017*, 4(1).

22. M. S. Swetha, S. K. Pushpa, M. S. Muneshwara, T. N. Manjunath, "Blockchain enabled secure healthcare Systems", *2020 IEEE International Conference on Machine Learning and Applied Network Technologies (ICMLANT)*, Hyderabad, India.

23. R. Anand, M. Pushpalatha, R. M. Patil, (2016). A social networking for sharing infrastucture resources in the social cloud computing. *International Journal of Informative & Futuristic Research* (IJIFR).

24. M. S. Muneshwara, M. S. Swetha, M. Thungamani, G. N. Anil, "Digital genomics to build a smart franchise in real time applications", *2017 International Conference on Circuit, Power and Computing Technologies (ICCPCT)*, 2017, Kollam, India.

25. R. Anand, Priyanka, R. M. Patil, (2017). Health monitoring in aerospace system. *International Journal of Informative & Futuristic Research (IJIFR)*.

26. M. V. Vijaykumar, P. Jagadish, K. Shryavani, R. Anand, (2016). Authorized deduplication in hybrid cloud. *IJCSN International Journal of Computer Science and Network*, 5(3), 559–563.

Index

289

Printed in the United States
by Baker & Taylor Publisher Services